FEMINISM/POSTMODERNISM/ DEVELOPMENT

Feminism/Postmodernism/Development takes current postmodernist critiques of modernity, postmodern feminist concerns with representation of Third World women and the possibilities for postmodern feminist political action one step further by emphasizing their intersections and exploring new directions and themes.

Drawing on the experiences of "Third World" women and "women of color," this collection challenges the ongoing reliance on dualities and explores the new issues, "voices," and dilemmas in development theory and practice. The book identifies various parallel processes affecting minority and Third World women, resulting in negative representations and silencing of their development expertise in favor of the supposed "expert" knowledge of Western development specialists. Using case studies of women in Africa, Latin America and Asia, as well as women of "color," the collection suggests the gap between local development knowledge and Western development expertise can be (and is sometimes) bridged in practice. The concern is to challenge the "Orientalist" representations of Third World and minority women as well as the silencing of their development expertise, by exploring how development theory and practice can be transformed to reflect their experiences, knowledges and political mobilizations.

Feminism/Postmodernism/Development brings postmodern questions to the field of gender and development, and not only acknowledges the importance of Third World and minority women's experiences in development issues, but also attempts to identify conditions for a more open and inclusive approach to gender and development.

Marianne H. Marchand is a Senior Lecturer in the Department of International Relations and Public International Law at the University of Amsterdam.

Jane L. Parpart is Professor of History, Women's Studies and International Development Studies at Dalhousie University in Halifax, Canada.

INTERNATIONAL STUDIES OF WOMEN AND PLACE

Edited by Janet Momsen, *University of California at Davis*
and Janice Monk, *University of Arizona*

The Routledge series of *International Studies of Women and Place* describes the diversity and complexity of women's experience around the world, working across different geographies to explore the processes which underlie the construction of gender and the life-worlds of women.

Other titles in this series:

FULL CIRCLES
Geographies of women over the life course
Edited by Cindi Katz and Janice Monk

'VIVA'
Women and popular protest in Latin America
Edited by Sarah A. Radcliffe and Sallie Westwood

DIFFERENT PLACES, DIFFERENT VOICES
Gender and development in Africa, Asia and Latin America
Edited by Janet H. Momsen & Vivian Kinnard

SERVICING THE MIDDLE CLASSES
Class, gender and waged domestic labour in contemporary Britain
Nicky Gregson and Michelle Lowe

WOMEN'S VOICES FROM THE RAINFOREST
Janet Gabriel Townsend

GENDER, WORK, AND SPACE
Susan Hanson and Geraldine Pratt

WOMEN AND THE ISRAELI OCCUPATION
Edited by Tamar Mayer

FEMINISM/ POSTMODERNISM/ DEVELOPMENT

Edited by Marianne H. Marchand and Jane L. Parpart

London and New York

First published 1995
by Routledge
11 New Fetter Lane, London EC4P 4EE

Simultaneously published in the USA and Canada
by Routledge
29 West 35th Street, New York, NY 10001

Typeset in Baskerville by
J&L Composition Ltd, Filey, North Yorkshire

Printed and bound in Great Britain by
Biddles Ltd, Guildford and King's Lynn

British Library Cataloguing in Publication Data
A catalogue record for this book is available from the British Library

Library of Congress Cataloguing in Publication Data
Feminism/postmodernism/development/edited by Marianne H. Marchand
and Jane L. Parpart.
p.   cm.—(Routledge international studies of women and
place)
Includes bibliographical references and index.
1. Women in development.   2. Women in development—Case studies.
3. Women—Developing countries—Social conditions.   4. Feminist
theory.   5. Postmodernism.   I. Marchand, Marianne H.
II. Parpart, Jane L.   III. Series: Routledge international studies
of women and place series.
HQ1240.F46   1995
305.42—dc20   94–23866

ISBN 0–415–10523–4   0–415–10524–2 (pbk)

CONTENTS

v

ACKNOWLEDGEMENTS

This book finds its "origins" in a conference panel on "feminism, postmodernism and development" at the Vancouver-meeting of the International Studies Association. The overwhelmingly positive response to the panel convinced the editors that a book on the subject was long overdue. And thus a new book-project was born! The project/ book is (and tries to be) different in many respects. We have, for instance, attempted to set up a dialogue around certain preconceived notions about women and development. We also wanted the book's structure to reflect our concern to avoid premature closure on the topic. Interestingly, the project turned out to be a "postmodern" experience in and of itself. Editorial meetings, reflecting our busy travel schedules and the unanticipated transatlantic move of one of the editors, were held in various conference locations and in airport departure halls between connecting flights. These meetings certainly evoke images of dislocation, de-centering and the (academic) nomad.

The introduction and structure of this book has benefited from discussions with many people. We thank the members of the Feminist Theory and Gender Studies section of the International Studies Association for providing the supportive environment in which this project could come to fruition. The Association for Women in Development also hosted one of our panels, and the heated discussions testified once again to the relevance and contentious nature of the subject.

Jane Parpart's exploration of postmodernism, feminism and development owes much to Christine Sylvester, who challenged her to write a paper on the subject for the Vancouver meeting. Christine has remained an inspiration, along with many other members of the ISA Feminist Theory and Gender Studies section. I have also benefited from comments by Linzi Manicom, Lourdes Beneria, Caroline Moser, Shelly Feldman, Moira Lloyd, Victor Li, the editors of *Development and Change*, and Eva Rathgeber. Special thanks go to my colleagues Pat Connelly, Tania Li and Martha MacDonald, co-members of the Summer Institute on Gender and Development team in Halifax, whose skepticism has continually tested my thinking and my resolve. Finally, I wish to thank Marianne, my co-editor, whose friendship and collegiality have enriched both the book and my life.

Marianne Marchand's thinking on these issues has been deeply influenced by (still continuing) discussions with Cindy Weber, Mridula Udayagiri and Roxanne Doty. I am also indebted to the faculty members, in particular Kate Sonderegger and Chela Andreu, of the Women's Studies Program at Middlebury College for the supportive environment they created and for the discussions we had on postcolonial, feminist and postmodernist theory. Finally, I wish to extend a special thanks to my co-editor Jane Parpart for her suggestions, ideas, support and friendship.

This book has benefited greatly from the insightful comments and suggestions of the two series editors, Jan Momsen and Janice Monk as well as Routledge editor Tristan Palmer. All three took the time to offer thoughtful and timely criticisms, which, especially at the final stages, have added immeasurably to the cohesiveness and quality of the manuscript.

Finally, we dedicate this book to our husbands Tim Shaw and Steve Stusek, who have patiently endured our seemingly endless editing/discussions sessions, the weeks, indeed years, of preoccupied work on the manuscript and the frantic transatlantic phone and fax messages needed to wrap up the project. They insisted on some balance (get a life) and helped us survive the rigors of editing such a disparate and challenging collection.

NOTES ON CONTRIBUTORS

Suresht R. Bald is a Professor of Political Science at Willamette University in Oregon. She has written extensively on South Asian migrants in Britain.

Eudine Barriteau is Lecturer and Head of the Centre for Gender and Development Studies at the University of the West Indies, Cave Hill, Barbados. She has been involved in research, administration and coordination of regional projects in the Caribbean. Currently she is writing on gender and development planning in the postcolonial Caribbean. [Ed. Earlier works by this author in the Bibliography are cited as Eudine Barriteau Foster.]

Geeta Chowdhry is Associate Professor of Political Science at Northern Arizona State University. She has written extensively on development, gender and the international political economy. She is the author of *International Financial Institutions, the State and Women Farmers in the Third World* (forthcoming).

Julia V. Emberley is Assistant Professor of Women's Studies at the University of Northern British Columbia. She is the author of *Thresholds of Difference: Feminist Critique, Native Women's Writings, Postcolonial Theory.* Her recent research interest is a study in the material culture of fur and the interconnected histories of sexuality and imperialism.

Mitu Hirshman is a PhD candidate in sociology at the Graduate School of the City University of New York. She is working on her dissertation on feminist theory and development in the Third World. She is also working on a study of "The Orientalism and Masculinism of Women in Development: The 'Labor' Question."

Marianne H. Marchand is Senior Lecturer in International Political Economy, and Gender and Development, at the University of Amsterdam. She has contributed to Stubbs and Underhill, *Political Economy and the Changing Global Order* (1994), and the special *Third World Quarterly* issue on The South in the New World (Dis)order. She is currently working on gender, regionalism and development in Latin America.

Maria Nzomo is Senior Lecturer in Government and International Studies at the University of Nairobi (Kenya). She is involved in the research and publishing activities of AAWORD, ACTS (the African Centre for Technology and Science) and DAWN. She has played a leading role in the women's democracy movement in Kenya and is currently involved in grassroots activism as well as the National Coalition of Women.

Jane L. Parpart is Professor of History and Women's Studies as well as coordinator of International Development Studies at Dalhousie University. She has published a number of collections on women in Africa, including *Women and the State in Africa* with Kathy Staudt and *Patriarchy and Class: African Women in the Home and the Workplace* (1988) and *Women, Employment and the Family in the International Division of Labour* (1990) with Sharon Stichter. She is currently involved in a cooperative project with the University of the West Indies and the Summer Institute on Gender and Development (Halifax) to write a text on gender and development theory. She is also involved in gender and development training in Africa and Asia.

Catherine Raissiguier is currently Associate Director/Assistant Professor at the Center for Women's Studies at the University of Cincinnati. She is the author of *Becoming Women/Becoming Workers: Identity Formation in a French Vocational School* (1994).

Eva M. Rathgeber is based in Nairobi, Kenya where she is Regional Director for the Eastern and Southern Africa Office of the International Development Research Centre. From 1987 to 1992 she established and was first coordinator of IDRC's Gender and Development Unit. She holds a PhD from the State University of New York and an MA from McGill University. She has published widely on gender and development, science and technology policy and communications in developing countries. She is currently completing a book on Zimbabwe's agricultural research structure.

Christine Sylvester teaches in the Political Science Department and Women's Studies Program at Northern Arizona State University. She is the author of several articles on Zimbabwean politics, and of *Zimbabwe: The Terrain of Contradictory Development* (1991). She has also authored *Feminist Theory and International Relations in a Postmodern Era* (1994) and co-edited with Dennis Pirages *Transformations in the Global Political Economy* (1990).

Mridula Udayagiri was born and raised in south India. She worked in various urban community development projects in the cities of Bangalore and Mangalore after obtaining her master's degree in social work. She later worked as a researcher for a non-profit organization on policy-oriented research for women and children in rural development on behalf of bilateral and unilateral aid agencies and the government of India. She is currently a doctoral candidate in sociology at the University of California, Davis. Her dissertation research focuses on the political implications of the structural adjustment programs introduced in India in 1991. She lives in northern California.

ACRONYMS

AAWORD	Association of African Women for Research and Development
AFC	Agricultural Finance Corporation (Zimbabwe)
AWID	Association for Women in Development
AWRAN	Asian Women's Research and Action Network
CAFRA	Caribbean Association of Feminists for Research and Action
CEC	Commission for the European Community
CIDA	Canadian International Development Agency
CODESRIA	Council for the Development of Economic and Social Research in Africa (Dakar)
DAWN	Development Alternatives with Women for a New Era
EPZ	export processing zone
FAO	(UN) Food and Agriculture Organization
FTZ	free trade zone
GAD	Gender and Development
GMB	Grain Marketing Board (Zimbabwe)
GONGO	government non-governmental organization
IBRD	International Bank for Reconstruction and Development (World Bank)
IDRC	International Development Research Centre (Ottawa)
ILO	International Labor Organization
IMF	International Monetary Fund
INSTRAW	United Nations International Research and Training Institute for the Advancement of Women
ISER	Institute for Social and Economic Research, University of the West Indies
IWTC	International Women's Tribune Center (New York)
NGO	Non-governmental organization
NIC	Newly industrializing country
NORAD	The Royal Norwegian Ministry of Development Cooperation

OCCZIM	Organization of Collective Cooperatives of Zimbabwe
PAR	participatory action research
PHC	primary health care
SAP	structural adjustment program
SEDCO	Small Enterprises Development Corporation (Zimbabwe)
SEWA	Self-employed Women's Association (India)
SHIELD	Sustained Health Improvement through Expanded Livelihood Development
SIDA	Swedish International Development Agency
SPARC	Society for the Promotion of Area Resource Centers
TWW	Third World women (approach to GAD)
UN	United Nations
UNCED	United Nations Conference on Environment and Development
UNDP	United Nations Development Programme
UNESCO	United Nations Educational, Scientific and Cultural Organization
USAID	United States Agency for International Development
WAD	Women and Development
WAND	Women and Development Unit, University of the West Indies
WEDNET	Women, Environment and Development Network (Nairobi)
WIAD	Women in Agricultural Development
WICP	Women in the Caribbean Project
WID	Women In Development
WIDE	Women in Development Europe
WLSA	Women and the Law in Southern Africa Research Project
ZANU-PF	Zimbabwe African National Union–Patriotic Front, ruling political party in Zimbabwe

Part I

EXPLODING THE CANON: AN INTRODUCTION/ CONCLUSION

*Jane L. Parpart and
Marianne H. Marchand*

In the last few decades postmodernist critiques have increasingly domi-nated scholarship in the humanities and social sciences. Grand theories of the past have been called into question; particularities and differ-ence(s) have triumphed as universal claims to knowledge have come under fire. Feminist scholars have reacted to postmodernist thought in a number of ways. Some reject it outright, while others call for a synthesis of feminist and postmodernist approaches. This is particularly true of scholars concerned with the marginalization of Third World women and women of color in the North.[1] However, many scholars and activists concerned with development issues in the South, especially poverty and economic development, dismiss this approach as a "First World" preoccupation, if not indulgence, with little practical application for Third World women's development problems. Again other scholars have challenged this view, arguing for the relevance of postmodern feminist thought to development issues.

In order to explore rather than curtail this emerging debate, *Feminism/Postmodernism/Development* has brought together a number of specialists on gender and development issues and asked them to investigate the possibility that a more politicized and accessible version of postmoder-nist feminist thought could have relevance for the problems facing most women in the South and indeed many women in the North, particularly women of color. Contributors have been encouraged to challenge each others' assumptions and to explore the limitations and strengths of postmodernist feminist thought for gender and development theory

1

and practice. In contrast to other collections, we do not expect a synthesis to emerge from this exercise. Rather we have urged individual authors to write from their particular vantage point or locale. To remain true to the spirit of postmodernism, we have also decided against a conclusion, which, we felt, would bring premature closure to an ongoing debate. Instead we have opted for an extensive Introduction/ Conclusion which provides discursive space for debate and discussion within and beyond the confines of the book without imposing fixed answers to the questions posed.

POSTMODERNISM

Postmodernism is not easily encapsulated in one phrase or idea as it is actually an amalgam of often purposely ambiguous and fluid ideas. It represents the convergence of three distinct cultural trends. These include an attack on the austerity and functionalism of modern art; the philosophical attack on structuralism, spear-headed in the 1970s by poststructuralist scholars such as Jacques Derrida, Michel Foucault and Gilles Deleuze; and the economic theories of postindustrial society developed by sociologists such as Daniel Bell and Alain Touraine (Callinicos 1989). These various strands were first woven together under the rubric of "post-modernism" by Jean-François Lyotard, in his book *The Postmodern Condition*, where he summarized postmodernism as above all maintaining "an incredulity toward metanarratives" (1984: xxiii–iv, 5). Postmodernists, he argues, question the assumptions of the modern age, particularly the belief that rational thought and technological innovation can guarantee progress and enlightenment to humanity. They doubt the ability of thinkers from the West either to understand the world or to prescribe solutions for it. The grand theories of the past, whether liberal or Marxist, have been dismissed as products of an age when Europeans and North Americans mistakenly believed in their own invincibility. The metanarratives of such thought are no longer seen as "truth," but simply as privileged discourses that deny and silence competing dissident voices.[2] The struggle for universalist knowledge has been abandoned. A search has begun for previously silenced voices, for the specificity and power of language(s) and their relation to knowledge, context and locality.

Michel Foucault, one of the leading postmodernist (and poststructuralist) thinkers, has emphasized the inadequacies of metanarratives and the need to examine the specificities of power and its relation to knowledge and language (discourse). He dismisses "reason" as a fiction and sees "truth" as simply a partial, localized version of "reality" transformed into a fixed form in the long process of history. He argues that discourse – a historically, socially and institutionally specific structure of

2

statements, terms, categories and beliefs – is the site where meanings are contested and power relations determined (Scott 1988: 36).[3] The ability to control knowledge and meaning, not only through writing but also through disciplinary and professional institutions, and in social relations, is the key to understanding and exercising power relations in society. According to Foucault, the false power of hegemonic knowledge can be challenged by counter-hegemonic discourses which offer alternative explanations of "reality" (Foucault 1972; 1979; 1980).

However, since no individual can comprehend the "truth," the limitations on knowing and on subjective experience have to be acknowledged. The search to understand the construction of social meanings has led postmodernist/poststructuralist scholars to recognize the contingent nature of the subject. As Judith Butler points out, "No subject is its own point of departure" (1992: 9). Individual subjects experience and "understand" life within a discursive and material context. This context, particularly the language/discourse that "explains" the concrete experiences of daily life, influences and shapes the way individuals interpret "reality." The self is thus not simply a reflection of experience (i.e. "reality"); it is constituted in complex historical circumstances that must be analyzed and understood as such. This more nuanced approach to the subject does not deny agency (the ability to act). "It is not a repudiation of the subject, but, rather a way of interrogating its construction as a pregiven or foundationalist premise" (Butler 1992: 9).

The contingent and constructed nature of the subject has drawn attention to the power of language/discourse and its impact on the way people understand and assign meaning to their lives. It has led to a call for the dismantling or deconstruction of language/discourse in order to discover the way meaning is constructed and used. Jacques Derrida (1976) in particular emphasizes the crucial role played by binary opposites. Indeed, he argues that Western philosophy largely rests on opposites, such as truth/falsity, unity/diversity or man/woman, whereby the nature and primacy of the first term depends on the definition of its opposite (other) and whereby the first term is also superior to the second. These pairs are as embedded in the definition of their opposite as they are in the nature of the object being defined, and they shape our understanding in complex and often unrecognized ways. In order better to understand this process, Derrida and others have called for the critical deconstruction of texts (both written and oral) and greater attention to the way difference(s), particularly those embedded in binary thinking, are constructed and maintained (Culler 1982).

These insights have spawned an interest in the construction of identity and the concept of difference(s). The search to discover the way social meanings are constructed has highlighted the importance of difference and the tendency for people to define those they see as different ("other") in opposition to their own perceived strengths. European and North American scholarship, benefitting from its hegemonic position in world discourse, has of course dominated the construction of such definitions. For example, Western Orientalists have for the most part defined and represented the Orient as inherently irrational and unreliable. This construction has projected the dark side of the West on to the Oriental "other," thus reinforcing the superiority of the supposedly rational, scientific West (Said 1979: 9–10). This focus on the hegemonic nature of colonial/neo-colonial discourse has shifted of late. Scholars such as Sara Suleri call for a more interactive approach, one that recognizes the interplay between those who "control" discourse and those who "resist" the dominant discourse. Colonial and neo-colonial discourses thus do not merely reflect the construction of a Third World "other," they also are influenced by postcolonial voices/discourses as well (1992: 4–7).[4]

In sum, postmodernist thinkers reject universal, simplified definitions of social phenomena, which, they argue, essentialize reality and fail to reveal the complexity of life as a lived experience. Drawing on this critique, postmodernists have rejected the search for broad generalizations. They emphasize the need for local, specific and historically informed analysis, carefully grounded in both spatial and cultural contexts. Above all, they call for the recognition and celebration of difference(s), the importance of encouraging the recovery of previously silenced voices and an acceptance of the partial nature of all knowledge claims and thus the limits of knowing.

POSTMODERNISM/FEMINISM

Feminists have responded to postmodern ideas in a number of ways. The strongest opposition has come from feminists working in the liberal (modern) or Marxist traditions, both of which are embedded in Enlightenment thinking. Liberal feminists, who have been preoccupied with policy formulation and the improvement of women's status within the structures of Western thought and society, generally write as if postmodern critiques have little or no applicability for their own work. This is particularly true of the many reports churned out on the status of women by established institutions, especially universities, government bureaucracies and international agencies such as the United Nations (UN), the World Bank (IBRD) and the International Labor Organization (ILO) reports on the status of women (see also Gillespie 1989; Joekes 1987). The possibility that "modernization" and "progress" may be unobtainable and undesirable goals in a postmodern world has rarely been considered by liberals working within these structures (World Bank 1979a; 1989b; 1993).[5]

Marxist feminists have also expressed considerable opposition to postmodern ideas. Sylvia Walby argues that "postmodernism in social theory has led to the fragmentation of the concepts of sex, race and class and to the denial of the pertinence of overarching theories of patriarchy, racism and capitalism." The postmodern critique of grand theory "is a denial of significant structuring of power, and leads to mere empiricism" (1990: 2). While more guarded in her assessment of postmodernism, particularly the focus on difference, Nancy Hartsock's analysis of Foucault leads her to the conclusion that, despite his stated preference for resistant discourse, his thought is deeply embedded in the dominant perspective. As a result, while he has much to tell us about the individual perception and experience of power, "systematically unequal relations of power ultimately vanish from Foucault's account of power" (1990: 165; see also Dubois 1991). Foucault's world is "a world in which passivity or refusal represent the only possible choices. Resistance rather than transformation dominates his thinking and consequently limits his politics" (Hartsock 1990: 167). This is a position Hartsock refuses to adopt.[6]

Some feminists believe feminist theory has always dealt with postmodern issues and indeed, has more to offer women than male-centric postmodern writers. Feminist anthropologists, Frances Mascia-Lees, Patricia Sharpe and Colleen Cohen (1989), attack postmodern anthropology for its profoundly sexist nature, noting that studies such as George Marcus and Michael Fischer's *Anthropology as Cultural Critique*, ignore feminist contributions to the discussion of the "other" and long-standing feminist critiques of Western notions of "truth." These scholars see postmodernist anthropology as an attempt to stave off the loss of Western male power "by questioning the basis of the truths that they are losing the privilege to define" (1989: 401–2). They believe feminism, with its openly political stance and its grounding in actual differences among women, has more to offer anthropology and the search for sexual justice than postmodernist theory (see also Tress 1988).

Similar arguments have been launched by the standpoint feminists, such as Sandra Harding (1987; 1992), Somer Brodribb (1992) and Dorothy Smith (1990), who focus on women's lived experiences as the basis of feminist knowledge. This "feminine" knowledge, according to Smith, "disrupts and disorganizes the discourse of modernity" (1990: 203), and thus contributes to the postmodern assault on modernity. However, standpoint feminists reject postmodernist attacks on the subject, as their critique of male hegemony is based on the authority of a female subjectivity grounded in women's daily lives. These scholars see no need, and indeed considerable danger, in adopting male-centric postmodernist thinking, particularly the attack on the subject. As Brodribb puts it: "As for the idea that feminists should be ragpickers

5

in the bins of male ideas, we are not as naked as that. . . . The very up-to-date products of male culture are abundant and cheap . . . what we have is a difficulty in refusing, of not choosing masculine theoretical products" (1992: xxiii).[7]

Feminists of various persuasions have expressed concern about the political implications of a postmodernist feminist perspective. Linda Hutcheon, for example, believes postmodernism threatens feminism's transformative agenda. She is particularly wary of postmodernist scepticism towards the subject. "Postmodernism," she argues, "has not theorized agency; it has no strategies of resistance that would correspond to the feminist ones" (1989: 168). Linda Alcoff reiterates this theme, pointing out that "If gender is simply a social construct, the need and even the possibility of a feminist politics becomes immediately problematic. What can we demand in the name of women if 'women' do not exist and demands in their name simply reinforce the myth that they do?" (1988: 420). The postmodernist focus on difference worries many feminists as well. As Susan Bordo points out, overemphasis on difference can lead to political fragmentation and the dissipation of feminist consciousness and activism. Indeed, she believes "feminism stands less in danger of the 'totalizing' tendencies of feminists than of an increasingly paralyzing anxiety over falling (from what grace?) into ethnocentrism or 'essentialism'" (Bordo 1990: 142; see also Di Stefano 1990a; Hawkesworth 1989).[8]

However, while agreeing that postmodernism, taken to its extreme and as practiced by its mostly white middle-class male proponents, appears to undermine feminists' search for a more egalitarian world, a growing number of feminists believe postmodernist thinking has much to offer feminist theorizing and action. Some see little conflict between postmodern thought and feminist politics (Butler and Scott 1992; Flax 1992; Hekman 1990). Others are more sceptical about postmodernist ideas and call for "an encounter, a strategic engagement between feminism and poststructuralism [postmodernism], that transforms both sides in significant ways" (Canning 1994: 373; Chow 1992; Fraser and Nicholson 1990; Lloyd 1991).

One of the most appealing aspects of postmodernism to many feminists has been its focus on difference. The notion that women have been created and defined as "other" by men has long been argued and explored by feminists in the North, most notably Simone de Beauvoir in her book *The Second Sex* (1949). She challenged male definitions of "woman" and called on women to define themselves outside the male/female dyad. Women, she urged, must be the subject rather than the object (other) of analysis. This concern was echoed and enlarged by other feminists, particularly those calling for the recovery of women's voices and the development of knowledge from the standpoint of women (Harding 1987; 1989).

However, the concern with women as "other" emanated largely from the writings of middle-class white women from Europe and North America, whose generalizations were grounded for the most part in their own experiences. Feminist theory produced by these scholars generally "explained" women as if their reality applied to women from all classes, races, cultures and regions of the world. Feminist concern with female "otherness" ignored the possibility of differences among women themselves (Gilligan 1982; Spelman 1990).

Not surprisingly, the postmodern focus on difference coincided with a growing critique of this position and thus provided further ammunition to women who felt excluded by the writings and preoccupations of these scholars and activists. Black and Native women in North America became increasingly vocal about their unique problems, arguing that race, culture and class must be incorporated into feminist analysis. While minority feminists have been calling for some time for racially and ethnically specific feminisms (Anthias and Yuval-Davis 1990; Anzaldua 1990; Collins 1989; Lorde 1981; 1984; Moraga and Anzaldua 1981), postmodernism has provided a space which legitimizes the search for "the voices of displaced, marginalized, exploited and oppressed black people" (hooks 1984: 25; 1992). Bell hooks eloquently argues for an African–American postmodernism where difference and otherness can be used to explore the realities of the black experience in North America and the connection between that experience and critical thinking. Only then, she argues, will feminism truly incorporate difference into its analysis (1984; 1990; see also Collins 1989; 1990; James and Busia 1993). Julia Emberley (1993) makes a similar argument for Native Canadian women.

A number of feminists in the South have taken up this position as well. They have accused Northern scholars of creating a colonial/neo-colonial discourse which represents women in the South as an undifferentiated "other," oppressed by both gender and Third World underdevelopment. Chandra Mohanty analyzes the writings about Third World women by a number of Northern feminists and discovered that women in the South have been represented as uniformly poor, powerless and vulnerable, while women in the North remain the referent point for modern, educated, sexually liberated womanhood. This analysis both distorts women's multiple realities and reduces the possibility of coalitions among (usually white) feminists from the North and working-class and feminist women of color around the world (1991a; 1991b). Or, as Aihwa Ong puts it: "For feminists looking overseas, the non-feminist Other is not so much patriarchy as the non-Western woman" (Ong 1988: 80; see also Barriteau Foster 1992; Chow 1992; and chapters 1, 3 and 9).[9]

7

The tendency to essentialize and distort the lives of Third World women does not just occur in the writings of women in the North. It is also pronounced in some of the work of Third World scholars trained in Northern institutions, particularly when writing for a Northern audience. As Rey Chow points out, Orientalists are not only white, they can also be non-Western students from privileged backgrounds "who conform behaviorally in every respect with the elitism of their social origins . . . but who nonetheless proclaim dedication to 'vindicating the subalterns'" (1993: 14). Similarly, Marnia Lazreg discovered that writings on Algeria by both Northerners and indigenous scholars trained in the North generally reflect Northern stereotypes about Arab peoples and culture, particularly the primacy of Islam, which is seen as a self-contained and flawed belief system impervious to change. She calls for a new approach, one that challenges prevailing (Western) paradigms while also revealing Third World women's lives "as meaningful, coherent and understandable instead of being infused 'by us' with doom and sorrow" (1988: 98; see also Schick 1990; Spivak 1990; and chapter 1).

These critiques have inspired considerable soul-searching among feminists in the North, and have encouraged an openness to difference and a reluctance to essentialize "woman" that bodes well for global feminist understanding. The writings emerging in this vein draw heavily on postmodernist thinking, particularly the focus on language and the question of subjugated knowledge(s) (Eisenstein and Jardine 1988; Riley 1988; Spelman 1990). They reveal a sensitivity to historical, spatial and cultural specificity (Momsen 1991; Weedon 1987), a recognition of the multiple oppressions of race, class and gender (hooks 1990; James and Busia 1993), a focus on the body as a locus of social control (Jaggar and Bordo 1989) and a commitment to uncovering previously ignored voices and resistances (di Leonardo 1991; Hirsch and Keller 1990). The emphasis on the role of place and location in the construction of identities and difference(s), particularly the emphasis on marginality as a site of resistance, has aroused a new interest in the way spatial contexts influence women's lives (Momsen and Kinnaird 1993; Pratt and Hanson 1994).

While few feminists argue for the wholesale adoption of postmodernist thought, and most continue to worry about its political implications, feminists of various persuasions are increasingly convinced that at least some aspects of postmodernist thinking are relevant to feminist theor(ies) and praxis. These syntheses have taken a number of forms. Some scholars believe postmodernist thinking can be readily incorporated into feminist theory and politics. For example, rather than accept Nancy Hartsock's assertion that Foucault undermines feminist politics, Joan Scott sees his pessimism:

as a warning against simple solutions to difficult problems, as advising human actors to think strategically and more self-consciously about the philosophical and political implications and programs they endorse. . . . Foucault's work provides an important way of thinking differently (and perhaps more creatively) about the politics of the contexual construction of social meanings, about such organizing principles for political action as 'equality' and 'difference'.

(1988: 36)

Judith Butler and Jane Flax see no reason why feminist politics requires a subject capable of "knowing" the "truth." Indeed, the "truth" is no guarantee of justice, for it has been used to justify atrocities such as the Holocaust and the Gulf War. A postmodern scepticism towards the subject and universal truths neither repudiates the subject, nor precludes the possibility that feminists can make claims about and act upon injustice. Rather it requires feminists:

> to firmly situate ourselves within contingent and imperfect contexts, to acknowledge differential privileges of race, gender, geographic location, and sexual identities, and to resist the delusory and dangerous recurrent hope of redemption to a world not of our own making. We need to learn to make claims on our own and others' behalf and to listen to those which differ from ours, knowing that ultimately there is nothing that justifies them beyond each person's desire and need and the discursive practices in which these are developed, embedded and legitimated.

(Flax 1992: 460)

These scholars do not advocate a simple marriage between feminism and postmodernism, but they believe feminism's attempts to challenge male hegemony from within male-centric Enlightenment thoughts are doomed to failure. They call for an alliance between feminism and postmodernism, based on their common critique of the modernist episteme. A feminism armed with the discourse theory of knowledge, according to Hekman, can fashion "new discourses about the feminine . . . that resist the hegemony of male domination." This alliance need not require a critical reassessment of postmodernism itself (Hekman 1990: 189–90; Canning 1994).[10]

A number of feminists who are sympathetic to postmodernist thinking adopt a different approach. They call for a strategic engagement between feminist and postmodernist thought, but one which will transform both perspectives rather than simply seek an alliance between the two. Nancy Fraser and Linda Nicholson, who have been in the vanguard of this debate, emphasize the similarities between feminist and postmodernist thought. Both, they argue, have sought to develop new paradigms of social criticism which do not rely on traditional philosophical underpinnings. While postmodernism has focused on the philosophical side of the problem, feminists have been more concerned with political questions. Fraser and Nicholson believe the two approaches complement each other. "Postmodernists offer sophisticated and persuasive criticisms of foundationalism and essentialism, but their conceptions of social criticism tend to be anaemic. Feminists offer robust conceptions of social criticism, but they tend at times to lapse into foundationalism and essentialism." They call for a critical engagement between the two, one that combines "a postmodernist incredulity toward metanarratives with the social-critical power of feminism" (Fraser and Nicholson 1990: 20, 34).

Much current feminist writing is concerned with this debate.[11] In *Materialist Feminism and the Politics of Discourse*, Rosemary Hennessy calls for a feminism that recognizes the importance of difference(s) and local complexities without abandoning attention to larger political and economic structures. She believes feminists should adopt "a way of thinking about the relationship between language and subjectivity that can explain their connection to other aspects of material life" (1993: 37). Rey Chow (1992: 114) advocates the "careful rejection of postmodernist abandon" while remaining committed to the need for the carefully situated, historical analysis of difference. Mary Poovey urges feminists to rethink "deconstruction" in order to make it useful for feminist agendas (1988: 51, 60–63).

The issue of subjectivity, and the limitations that postmodern critiques of the subject pose for the female voice/subject, are of particular concern to many feminists otherwise sympathetic to postmodernist thinking. Loath to lose the authority of the female voice, yet wary of the unproblematized subject, some feminists are calling for a female subjectivity characterized by partial identities and mobile subjectivities (K.E. Ferguson 1993: 158; see also Goetz 1991). They argue for a "reconceptualization of the subject as shifting and multiply organized across variable axes of difference" (de Lauretis 1990: 116). These multiple, mobile subjectivities are embedded in the historical, spatial and institutional contexts of daily life and must be understood in that context (Canning 1994). This approach integrates the standpoint feminists' "effort to interpret the subject women, and the postmodernist effort to examine how specific subjects came to be (or not) and what they have to say. . . . It looks for new forms and mobilities of subjectivity that can replace single-subject categories, inherited homes, without denying, nonrecognizing, the currently existing subject" (Sylvester 1994: 59).

For these scholars, and many others, the encounter between feminism and postmodernism is clearly ongoing, indeterminate and fluid. It is contested terrain, which will no doubt continue to foster debate and dialogue. For many feminists, it provides an arena where differences and ambiguities can be celebrated without sacrificing the search for a "broader, richer, more complex, and multilayered feminist solidarity, the sort of solidarity which is essential for overcoming the oppression of women in its 'endless variety and monotonous similarity'" (Fraser and Nicholson 1990: 35).

RETHINKING DEVELOPMENT AND GENDER

Does this debate have anything to offer theorists and practitioners concerned with development, particularly women's development? We think it does. The critique of modernity and Western hegemony, the focus on difference and identity, the emphasis on the relationship between language and power, the attention to subjugated knowledge(s) and the deconstruction of colonial and postcolonial representations of the South as the dependent "other" have much to say to those involved in the development business.

After all, the development enterprise, whether drawing on liberal or Marxist perspectives, has been largely rooted in Enlightenment thought. The liberal approach to development grew out of the postwar period in the 1940s, when economists and policy makers believed development could be achieved by the simple adoption of Western political and economic systems. The Marxists, while critical of international capital and the class structure, never questioned the equation between modernization and development. Both saw development as a fairly straightforward, linear process, in which a nation or people moved from underdevelopment, which was equated with traditional institutions and values, to full development, i.e. modern/rational/industrialized societies based on the Northern model (Johnston 1991). The rationale for this progression was provided by colonial (and later neo-colonial) discourses which compared "backward, primitive" Third World peoples and cultures unfavourably with the "progressive" North (Curtin 1974; Said 1979; 1993). More recently, economic crises in the South and the demise of socialism have reinforced a predominantly American version of development, which draws on neoclassical economics, modernization theory and evolutionism to provide "failsafe" prescriptions for Third World development (often known as structural adjustment programs (SAPs)) (Pieterse 1991). However, development has continued to be seen largely as a logistical problem. Its goal, to make the world modern, i.e. Western, has rarely been in dispute.

This approach has not gone unchallenged. In the 1960s, continuing poverty in the South inspired a critique of mainstream development. Latin American scholars argued that international capitalism, far from

11

developing the world, was perpetuating and benefiting from Third World underdevelopment (Frank 1978). They called for separation from the centers of international capital (the metropole) and for self-reliant development in the South (periphery). This attempt to turn modernization on its head soon ran into trouble. Critics questioned the linear, inevitable characterization of metropole/periphery relations, and the lack of attention to class forces in the Third World (Cooper 1981). While debates within this perspective still continue (Munck 1993), a number of scholars on the left have responded to the current impasse in development theory by adopting a post-Marxist approach to development.[12] Acknowledging the limitations of classical Marxist analysis, particularly its economistic, linear character, these scholars emphasize instead the fluid, contingent nature of capitalist development, the importance of human agency and the complexity of social transformation (Booth 1985; Corbridge 1989; Schuurman 1993a and b; Slater 1992).

While this is an important and still developing critique, it has done little to undermine the equation between development, modernity and the West. In contrast, scholars drawing on postmodernist perspectives have challenged the very essence of mainstream and radical development discourse (DuBois 1991; Escobar 1984–5; 1992; Ferguson 1990; Johnston 1991; Moore 1992; Pieterse 1991; Slater 1993). Above all, they question the universal pretentions of modernity, and the Eurocentric certainty of both liberal and Marxist development studies. They point out that much of the discourse and practice of development has exaggerated Western knowledge claims, dismissed and silenced knowledge from the South and perpetuated dependence on Northern "expertise." They call for a new approach to development, one that acknowledges difference(s), searches out previously silenced voices/knowledges and recognizes the need to welcome multiple interpretations and "solutions" to developmental problems. While some scholars argue for a synthesis of post-Marxist and postmodern approaches (Schuurman 1993a; Slater 1992), all celebrate postmodernism's "emphasis on iconoclastic questioning rather than predetermination, on openness rather than pre-empting closure, on plurality rather than essentialism" (Slater 1992: 311).

The impasse within development theory has been exacerbated/aggravated by recent changes in the world's geopolitical outlook. The crumbling of the Soviet empire was interpreted by Francis Fukuyama (1989) as a victory of (Western) liberalism and, by extension, of modernization theory. Given the ongoing processes of global political and economic restructuring which are also seriously affecting the "former West," the issue is more complicated than Fukuyama believed. Within the field of development studies a serious rethinking of the embeddedness of modernization theory in Cold War (ideological) structures is necessary.[13] In addition to shedding its Cold War legacy, the field of development studies also needs to address the changing global realities, including the restructuring of the international political economy as well as, quite paradoxically, the emergence of regionalism. New categories of possible recipient countries and groups are emerging, including some people in the North. Clearly, this renaming/relabeling requires students of development to critically analyze the current restructuring of development practices.[14]

Are similar issues and critiques relevant for the development of women? We think they are. Certainly development specialists initially saw Third World women as an impediment to development, if they considered them at all. They readily adopted colonial representations of Third World women "as exotic specimens, as oppressed victims, as sex objects or as the most ignorant and backward members of 'backward' societies" (de Groot 1991: 115). This vision of Third World women as tradition-bound beings, either unable or unwilling to enter the modern world, fit neatly into Western and neo-colonial gender stereotypes, and provided a rationale for ignoring women during the first two development decades (1950s–60s). WID

By the mid-1960s some economists began to realize that development was not taking place as easily as they had hoped, particularly in regard to women. In 1970 Ester Boserup's landmark study, *Woman's Role in Economic Development*, reported that many development projects, rather than improving the lives of Third World women, had deprived them of economic opportunities and status. Inspired by Boserup's work, a new subfield of development, Women in Development (WID) gradually emerged.[15] Steeped in the liberal tradition, this approach sought greater equity between women and men, but Western gender stereotypes went largely unchallenged. Women's development was seen as a logistical problem, rather than something requiring a fundamental reassessment of gender relations and ideology (Rathgeber 1990; Tinker 1990). WAD

In the 1970s, emerging critiques of development and patriarchy influenced some development practitioners concerned with women's issues. The dependency theorists' call for self-reliant development (Amin 1974; 1977) and the radical feminists' assertion that women could only develop outside patriarchal power structures (Daly 1978), served as a backdrop for a new approach to the development of women, one that built on and celebrated women's culture, emphasized women-only projects, and warned against close cooperation with male-dominated institutions. This approach, sometimes known as Women and Development (WAD), has been influential in the policy and program-

13

ming of many non-governmental organizations (NGOs) (Parpart 1989; Rathgeber 1990).[16]

WID proponents responded to these critiques by modifying mainstream development policy for women as well. Concern with equality between women and men fell by the wayside as planners emphasized basic human needs, particularly for health, education and training. WID specialists argued that this approach would increase women's effectiveness and efficiency at work, thus assisting both economic development and women's lives (Kandiyoti 1990; Moser 1993).

In the 1980s, some scholars and activists in both the North and South began calling for a new approach to women's development. Drawing on concerns about the growing poverty of women (and men) in the South, as well as radical feminist concern with global patriarchy, a socialist feminist position gradually emerged in various parts of the world. Feminists in the South contributed to this debate as they sought their own answers to developmental problems. The series of international conferences celebrating the UN Decade for Women (1976–85) highlighted the unique problems facing women in the South and encouraged the development of organizations to foster research and writing by Third World scholars.[17] The scholarship emerging from these organizations (AAWORD 1986; Sen and Grown 1987) has strengthened the voice(s) of Southern scholars and activists, and is providing the basis for feminist theorizing and action grounded in Southern realities (more recently called the "empowerment" approach to women's development (Moser 1993), see also chapters 1 and 12)). It has also inspired links with feminists in the North concerned with global and gender inequities. A commitment to understanding class, race and gender inequality in a global context thus provided an intellectual meeting-point for likeminded feminists from around the world.[18]

The resulting dialogue, increasingly known as Gender and Development (GAD), focuses on gender rather than women, particularly the social construction of gender roles and relations. "Gender is seen as the process by which individuals who are born into biological categories of male or female become the social categories of men and women through the acquisition of locally-defined attributes of masculinity and femininity" (Kabeer 1991: 11). The possibility of transforming gender roles is thus established, for the gendered division of labor and power is revealed as a constructed rather than natural part of life. While GAD proponents rarely challenge the goal of modernization/Westernization, some scholars believe the GAD perspective provides the possible (discursive) space to do so (see chapter 11).

This approach has had considerable influence on academic development discourse, but its willingness to consider fundamental social transformation does not sit well with most mainstream donor agencies. Most development agencies shy away from the language of transformation, particularly in such a sensitive area. Although some government agencies (most notably the Scandinavians, Dutch and Canadians) and some NGOs have officially adopted a more gender-oriented, transformative approach to their development programs, at least at the level of training, this perspective is rarely integrated into development planning (Moser 1993).[19] Development discourse, whether mainstream or radical, for the most part continues to characterize women (and children) in the South as vulnerable, helpless victims (Commonwealth Expert Group 1989; Cornia *et al.* 1987). Renewed interest in efficiency and increased donor support for women entrepreneurs (Moser 1993: 56–7) has done little to shake this image or to undermine most development specialists' belief in the modernization project.

In recent years various scholars have started to look at the intersection between gender, development and the environment. In the view of some authors, women's relation to nature is different (and more "natural") than that of men. Moreover, they emphasize that there are parallels/similarities between patriarchy (i.e. man's dominance over woman) and man's attempt to dominate nature. Therefore the best starting-point for formulating sustainable development strategies is (to analyze) the relationship between women and nature (Mies and Shiva 1993). While sympathetic to a woman-centered focus to environmental issues, scholars such as Bina Agarwal (1991) reject the essentialist character of this argument and call for a more nuanced, materialist approach. Others draw on feminist and postmodern critiques of science and modernity to challenge the growth model of development and its implications for women and the environment (Braidotti *et al.* 1994).

POSTMODERN FEMINISM, GENDER AND DEVELOPMENT

While these various perspectives have contributed important theoretical and practical insights to developmental issues facing women in the South (and many women in the North), they have for the most part ignored some of the more intractable impediments to women's development. They rarely challenge Northern hegemony, nor have they been able to provide the tools to dismantle patriarchal gender ideologies in the South (and North). This collection explores/challenges the possibility that a critical and flexible adoption of postmodern feminist thinking can provide the basis for a more sensitive, and transformative, approach to the development of women in the South (and North).

The issue of colonial/neo-colonial discourse has thus far been the most immediate "link" between postmodern feminism and gender and development. The critique of colonial constructions of the "Third World

15

woman" has revealed the hierarchical, dualistic nature of Western thought, and its tendency to reify and reinforce the North/South divide for women as well as men. While some scholars have acknowledged the interactive nature of colonial discourse, the emphasis has been on the discourse of the powerful and the hegemonic character of colonial/neo-colonial representation.

This book moves beyond the critique of colonial/neo-colonial discourse and representation to raise new issues and debates within the context of the gender and development literature. It argues for the contextualization of colonial/neo-colonial discourse. Who engages in it? Is it only Northern feminists and development practitioners who (mis)represent women in the South? What about Western-educated women (and men) in the South (and those living in the North) who speak for their oppressed sisters yet live in comfortable urban surroundings? What about the activists who emphasize the helplessness of Third World women in order to mobilize funds from foreign donors? Are there (dis)similarities between the representations of Third World women and minority women in the North?

The contributors agree that theorists and practitioners working within the WID perspective, whether from the North or the South, have tended to represent women in the South as the backward, vulnerable "other," in need of salvation from Northern (or Northern-trained) development experts. This language, whether expressed by Northern experts or by their allies in the South, has reinforced the myth of the North/South divide, with its assumption that the "developed" North controls the keys to technology and modernity required by the South. While individual exceptions occur, WID discourse has generally fostered development practices that ignore difference(s), indigenous knowledge(s) and local expertise while legitimating foreign "solutions" to women's problems in the South.

The contributors disagree on the character and consequences of representations by proponents of the GAD and empowerment perspectives. Some believe GAD analysis provides a space for a deeper understanding of Third World women's lives; others criticize its modernist tendencies. Some see writings from the South, particularly from DAWN (Development Alternatives with Women for a New Era), as the basis for a new approach to women's development. Others believe the emphasis on poor women has essentialized women in the South, obliterating difference(s) and reinforcing many of the negative stereotypes associated with the WID position (see chapters 2 and 11). Indeed, the literature on minority women in the North reveals many of the same tendencies. Clearly we need to interrogate *all* representations if we are to understand their impact on the self-perceptions and actions of women in the South (and North). The poor, vulnerable Southern woman is a powerful image, and its ready adoption by both mainstream and alternative development theorists and practitioners is understandable (Commonwealth Expert Group 1989; Sen and Grown 1987). Yet this very image reinforces and maintains the discourse of modernity so essential to Northern hegemony and development practices.

The need to deconstruct development discourse is all the more crucial at the current conjuncture, when global restructuring and environmental degradation are challenging the traditional realm being "covered" by gender and development theory (and practice). More and more there is talk about the Third World within the First World (see chapters 4–6). Increased differentiation in the South (as well as the new "South" in Eastern Europe and the former Soviet Union) is casting doubt on the myth of the North/South divide. Gender and development theory and practice has not been able adequately to incorporate/address these changes. It is still directed at the South, which partially serves to perpetuate the North's domination of the South. This position is increasingly untenable as emerging powers in the South, especially in Asia, and growing dislocations in the North, undermine old certainties and raise developmental issues for women in the North as well as the South. Clearly, new thinking about women/gender and development is required.[20]

The authors in this collection explore the possibility that postmodern feminist thinking can contribute to this task. As we have seen, the discourse of development has often disempowered poor women. This comes as no surprise to those who are critical of the dualistic, patriarchal language and assumptions embedded in Western development thinking. Rather than reject development all together, however, the contributors acknowledge the very real problems facing poor women in the South (and North), and the need to address developmental questions in an increasingly complex, interrelated and unequal world.[21] Postmodern feminist thinking, with its scepticism towards Western hegemony, particularly the assumption of a hierarchical North/South divide, provides new ways of thinking about women's development. It welcomes diversity, acknowledges previously subjugated voices and knowledge(s) and encourages dialogue between development practitioners and their "clients."

As several contributors point out (chapters 10–12), this approach brings the role of the development analyst/expert into question. It reminds us of the close connection between control over discourse/knowledge and assertions of power. Indeed, the authority of many Northern development "experts" and like-minded analysts in the South has been maintained and reinforced (until recently) by their virtual monopoly over the discourse on "Third World women." While acknowledging the individual contributions of many thoughtful and committed development practitioners, this critique raises important issues about development practitioners' claims to "know" the answers to women's developmental problems in the South (or North). It suggests the need for a more inclusive, open approach to women's development, one that rejects the tendency to undervalue women's knowledge unless it comes from Northern institutions or carries the seal of approval (i.e. certificates) from Northern experts.

The recovery of women's knowledge/voices, especially those of the poor, is not a simple matter however. The post-colonial literature, with its focus on the discourse of the powerful, offers important insights into the forces silencing women, but it has less to say about the way women actively construct their own identities within the material and discursive constraints of their lives. The postcolonial analysis of resistance, hybridity and mimicry, with its attention to the multiple forms of resistance against Western hegemony (Bhabha 1994; Spivak 1990), along with postmodern feminists' attention to difference, language and power, have much to offer discussion on this issue. The contributors grapple with the implications of this thinking for gender and development theory and practice. Above all, they emphasize the need to situate women's voices/experiences in the specific, historical, spatial and social contexts within which women live and work. However, they are wary of an unproblematic "Third World woman," and acknowledge the need to adopt an approach that recognizes the multiple axes/identities which shape women's lives, particularly race, class, age and culture. As Eudine Barriteau points out, these multiple identities/oppressions interact in complex and often unexpected ways, reminding us that women construct their lives in complicated and often shifting material and discursive environments which are both difficult to understand and to change (see chapters 4–6, 10).[22]

The celebration of differences and multiple identities has provided a welcome plurality and richness to feminist analysis, but in its extreme form, it raises questions about the ability of women to speak for each other, to mobilize political action and to call for the defense of women's rights (see chapters 7 and 9). If women's identities are constructed and fluid, and the world is full of uncertainty and confusion, how can women in both the South and North mobilize to defend their interests? Indeed, how can women even understand each other? And how can we deal with the recurrent theme of social/political/economic marginalization of most women? Postmodern feminists have drawn our attention to the dualistic, patriarchal tendencies in Western society and in much of the discourse on women's development. These deeply embedded structures of language and meaning interfere with communication. Dialogue across differences is not simply a matter of good will. It requires a recognition of the power of language, a conscious effort to situate one's own knowledge and a willingness to open oneself to different world views (Connelly *et al.*, forthcoming).

An approach to development that accepts and understands difference and the power of discourse, and that fosters open, consultative dialogue can empower women in the South to articulate their own needs and agendas. Instead of simply seeing women as a disempowered "vulnerable" group in need of salvation by Western expertise, Gender and Development experts can rethink their approach to development. Attention to women's lived realities and understandings, and genuine partnership between North and South can lead to development policies that foster self-reliance and self-esteem, rather than ignoring women's knowledge and creating policies and projects that increase patriarchal control over women's bodies and labor (Ferguson 1990; Johnston 1991). As a number of contributors point out, a postmodern perspective has much to offer both theorists and practitioners of development. It undermines Northern universalism, whether based on liberal or Marxist assumptions, and provides the intellectual basis for a new understanding of global diversity.[23]

However, there are dangers in this approach which must be addressed. Postmodern feminism, taken to its extreme, can stymie collective action among women, both within nations and on a world scale. The emphasis on difference and indeed on the often deep divisions between women in the South, minority women in the North and more privileged (often white) women in the North (and some in the South as well) offers both insights and dangers. While reminding us of the need to guard against glib assumptions about global feminist solidarity, this focus on difference can exacerbate differences among women and undermine possibilities for collective action by women, thus reinforcing the power of patriarchy and reducing the chance that women can challenge the gender hierarchies and ideologies that construct and maintain their subordination. Indeed, Maria Nzomo and Mridula Udayagiri question the ability of postmodern analysis to address these questions. As Maria Nzomo points out, the rejection of universals can undermine women's struggles for democratic rights and greater participation in development and the state. Others believe political decisions and action can be mobilized around an empathetic understanding of people's daily lives, based on the carefully situated, partial knowledge of people's daily lives rather than on empty slogans about "universal" principals (see Goetz 1991: chapters 3, 8, 10). The impenetrable jargon of much postmodern writings is an issue as well. Daunting even for the educated, postmodern language is often an insurmountable obstacle for people mired in widespread illiteracy and economic crisis.

The contributors have been encouraged to raise these issues, and others. We do not want premature closure of this important debate.

Rather we have challenged the contributors to think through and contest the issues at hand, in the hope that this debate will encourage the development of a more politicized and accessible version of postmodern feminist thought – one which can address the problems and prospects facing women in an increasingly complex and interrelated world.

NOTES

1 Throughout this book, the term Third World is adopted as a shorthand for Africa, Latin America and Asia, with the understanding that these areas, while exhibiting certain similarities, have many differences as well. The term is not to be seen as an assumption that Third World peoples, especially women, can be lumped in one undifferentiated category. Similarly, the term First World is a shorthand for the more industrialized countries, but it too suggests a uniform condition which is increasingly false, particularly for minorities. We have adopted the terms South and North to indicate our belief that global economic rivalries and status are no longer defined on an East–West axis, but rather increasingly around the less industrialized economies in the South and the more industrialized economies in both the North and South (particularly the NICs in Southeast Asia). The West refers specifically to Europe and North America.

2 Some Western scholars, most notably Marxists, reject postmodernism as dangerous and politically naive (Callinicos 1989; Palmer 1990). Others, while sympathetic to Marxism, see postmodernism as an outgrowth of the culture of late capitalism. Fredrick Jameson, for example, endorses an approach which draws on the strengths of postmodernism without abandoning political action (Jameson 1991). Some scholars find postmodernism's emphasis on difference and multiplicity useful for their work and not necessarily inimical to other approaches (Ankersmit 1990; Prakash 1990).

3 As Eagleton explains, "'Language' is speech or writing viewed 'objectively,' as a chain of signs without a subject. 'Discourse' means language grasped as utterance, as involving speaking and writing subjects and therefore also, at least potentially, readers and listeners" (Eagleton 1983: 115).

4 This book will use the terms colonial and neo-colonial discourse to refer to the writings/discourse of Western/Northern authors on the South. The term postcolonial refers to the writings of authors in the South, some of whom are based in the North. This terminology is fluid and a continuing matter for debate (see Bhabha 1994; McClintock 1992; Shohat 1992). The postcolonial critiques of Western discourse on the South has been particularly well developed among Indian historians (Guha and Spivak 1988; Prakash 1990).

5 Indeed, Sabrina Lovibond (1990: 179) argues that feminism "should persist in seeing itself as a component or offshoot of Enlightenment modernism, rather than as one more 'exciting' feature (or cluster of features) in a postmodern landscape." For reports from the World Bank, see the Staff Working Papers, the Country Studies and the Technical Papers.

6 Nancy Fraser shares many of Hartsock's misgivings, arguing that "Foucault calls too many different sorts of things power and simply leaves it at that . . . . What Foucault needs, and desperately, are normative criteria for distinguishing acceptable from unacceptable forms of power" (Fraser 1989: 32–3). Hartsock is often labeled a standpoint feminist as well.

7 Cultural feminism also stresses the primacy of women's experiences and culture, which is seen as an alternative to patriarchal ideologies and structures (Alcoff 1988: 408–14). Radical feminists such as Mary Daly (1978) and Adrienne Rich (1976) are influential proponents of this position.

8 Many feminists who are concerned about the negative political implications of postmodern analysis for feminist action still find Foucault's emphasis on contextual situated analysis useful (Diamond and Quinby 1988; McNay 1992; Rhode 1990; Sawicki 1991).

9 Also see Anzaldua 1990; Chow 1991b, 1993; Johnson-Odim 1991; Minh-ha 1989; Ong 1990, 1993; Sangari and Vaid 1989; Spivak 1990; and Torres 1991. Pauline Rosenau (1992: 152–5) speaks of these writers as Third World affirmative postmodernists.

10 Some feminists working closely within the postmodernist perspective are Joan Scott, Linda Singer, Jane Flax, Judith Butler and Susan Hekman.

11 Some of the many feminists seeking a strategic encounter between feminism and postmodernism are Kathleen Canning, Kathy Ferguson, Moya Lloyd, Chris Weedon, Donna Haraway and Christine Sylvester.

12 Schuurman blames the impasse in development theory on: the continuing gap between rich and poor; the lack of long-term planning in the 1980s; the realization that economic growth has had a catastrophic impact on the environment; the delegitimation of socialism as a viable alternative; the recognition that globalization undercuts the relevance of the nation-state; the growing differentiation within the Third World; and the advancement of postmodernism within the social sciences (1993b: 10–11, 1993c).

13 For example, Samuel Huntington, a Soviet specialist, has begun contributing to development studies (1993).

14 See *A World in Conflict* by the Dutch minister of development cooperation, Jan Pronk (1993), which assigns new labels to countries and regions.

15 Most development agencies did not have full-time WID professionals until the mid-1980s (Mueller 1987).

16 Although often relegated to small-scale, women-only projects, the WAD approach also influenced development practitioners working on international projects (see International Women's Tribune Center (IWTC) Newsletters).

17 Some examples are Development Alternatives with Women for a New Era (DAWN), which covers the entire Third World; the Association of African Women for Research and Development (AAWORD); the Women and Development Unit of the University of the West Indies (WAND); and the Asian Women's Research and Action Network (AWRAN).

18 Sussex University in England and the Institute for Social Studies at The Hague provided important meeting-points in Europe for like-minded scholars from around the world.

19 The GAD approach has offered development planners a way to differentiate between practical (i.e. specific, daily) gender needs and strategic (or more long term, empowerment) needs for women. This approach seems to be making some inroads into development thinking and planning, but primarily at the level of training (interview, Sherry Greaves, CIDA, WID Unit, Ottawa, 28 February 1992; also see Chow 1991a).

20 A first tentative attempt to capture these changes is the refocusing of Women in Development Europe (WIDE), a European network of development practitioners and scholars. WIDE is now starting to emphasize the need for a gender and ethnic analysis of the position of women in Europe in

order to better understand the impact of global restructuring on women in the North and South, as well as relations among them (WIDE Conference, Amsterdam, 28–9 May 1994).

21 Peet and Watts (1993: 238) point out that between 1960 and 1989, the top fifth of the world's population doubled its control over the world's riches, going from 30 to 60 percent. In contrast, the bottom fifth controls 1.4 percent.

22 Standpoint feminists have emphasized women as subjects, while postmodern feminists, black feminists and many Third World feminists have stressed the multiple, complex nature of women's subjectivity.

23 For example, the Chinese version of modernity, based on "a blending of capitalism with Chinese tradition and racial genius" (Ong 1993: 8), raises important issues for development theory and practice.

Part II

(NEO-)COLONIAL DISCOURSE(S) AND THE REPRESENTATION OF THIRD WORLD WOMEN AS THE "OTHER"

In this first part the authors, Geeta Chowdhry, Mitu Hirschman and Marianne Marchand, address the relevance of postmodernist thought for the Gender and Development literature. As is clear from the Introduction/Conclusion, postmodern feminism has challenged theorizing within the social sciences on various counts. In its critique of Enlightenment thinking, postmodernist feminism questions the claims to universalism and "truth" embedded in these theories. Various scholars have used postmodernist insights to criticize this literature for its embrace of colonial/neo-colonial discourse(s) portraying Third World women as tradition-bound, passive, voiceless and interchangeable objects (Lazreg, Mohanty, Ong). They argue that the production of knowledge (within the field of Gender and Development) is intricately related to the representation of Third World women as the "other." Generally agreeing with these criticisms, the contributors to this section go beyond this original claim and explore various related themes.

In her chapter Geeta Chowdhry explores the more general critiques of colonial discourse within the particular context of the World Bank's WID policies. According to her, the WID representation of Third World women encompasses three interrelated images: women in the South are shown as victims, sex objects and as "cloistered" beings, relegated to the private sphere or *zenana*. Within the WID policies of the World Bank the *zenana* representation is particularly dominant.

In addition to their embeddedness in colonial/neo-colonial discourse, WID policies are also enshrined in the liberal discourse on markets. According to Chowdhry, these two discourses together tend to disempower women, first by portraying them as passive, traditional objects and, second, by only focusing on their roles as producers. In other words,

23

the liberal discourse on markets ignores women's reproductive roles. Thus, WID policies are inherently contradictory and inconsistent: situating WID policies at the intersection between colonial/neo-colonial discourse and the liberal discourse on markets ultimately renders the stated objective of WID, i.e. to mainstream women in the development process, impossible.

It follows that an alternative development policy, which would reinforce or stimulate the empowerment of Third World women, cannot be found within the scope of these two discourses. According to Chowdhry, alternative strategies to WID have been formulated by Third World feminist scholars who draw on insights from both socialist and postmodernist feminisms, but who simultaneously ground their analyses in the material realities of women in the South. An examples of such a Third World empowerment perspective is the DAWN manifesto, *Development, Crises, and Alternative Visions* (Sen and Grown 1987).

Mitu Hirshman disagrees with Chowdhry's conclusions. Although she acknowledges that DAWN is an important departure from the WID perspective, she seriously questions DAWN's claim that it provides a "radical alternative to conventional development theory and practice, as well as to WID". In her critical reading of *Development, Crises, and Alternative Visions* she finds that DAWN has not been able entirely to divest itself of WID's assumptions, particularly regarding the "development question" and the "woman question." For one, Sen and Grown continue to use the sexual division of labor as the principal category of analysis. In Hirshman's view, the privileging of labor reveals a Western, modernist and male bias. Establishing women's labor as the lens through which to understand and analyze their experiences creates unnecessarily a hierarchy among different aspects of women's lived realities.

Likewise, by focusing primarily on *poor* Third World women, Sen and Grown still tend to cast these women as victims of the development process. In so doing, Hirshman argues, they continue to objectify (poor) Third World women. Only this time the intervention on the latter's behalf needs to come from their more educated sisters in the South. However, the problem of disempowerment continues to be present: (poor) Third World women are still denied agency.

Marianne Marchand, in turn, takes the discussion on the representation of women in the South one step further. In her chapter, she focuses on the connections between the representation of Latin American women and the silencing of their voices and knowledge(s). The subordination of, in particular poor, Latin American women's voices is realized through the distinction between feminist and feminine movements. This distinction is not only employed in Western feminist writings on Latin American women's movements, but also by Latin American women/feminists themselves. The distinction is based on the hierarchized

24

opposition between practical (or feminine) and strategic (or feminist) gender needs, a distinction which is widely used within the gender and development field (see Moser 1993). Through the elimination of this implicitly hierarchical dichotomy, a (discursive) space can be created. This provides the necessary opening(s) within the Gender and Development literature for Latin American women's theories and practices (as recorded in testimonies) to be heard and thus to contribute to the production of knowledge about Gender and Development in Latin America.

In sum, this first section on the intersections between the knowledge/power nexus and the representation of Third World women provides us with much food for thought. It is clear that trying to develop alternatives to mainstream development policies is a long and arduous process. It requires the reexamination of established orthodoxies, an open mind toward new insights and ideas, attention to language and a commitment to communication across diversities and differences.

1

ENGENDERING DEVELOPMENT?

Women in Development (WID) in international development regimes[1]

Geeta Chowdhry

This chapter will demonstrate that the so-called WID regime, as implemented by international development agencies, has its origin in two distinct yet overlapping strands of modernist discourse: the colonial discourse and the liberal discourse on markets. The colonial discourse based on the economic, political, social and cultural privileging of European peoples, homogenizes and essentializes the Third World and Third World women. The liberal discourse on markets, based on a negative view of freedom (Chowdhry 1993), promotes free markets, voluntary choices and individualism. Its epistemological premises and practical implementations disempower Third World nations in the international political economy. Moreover, as it intersects with colonial discourse, liberal discourse paradoxically tends to disempower poor Third World women (despite its stated objective of helping women to "develop"). In this chapter I argue that this disempowerment of Third World women is exemplified and embodied by the WID regime, because it is situated at the intersection of these two (modernist) discourses.

The chapter is organized into three main sections. The first section discusses the question of colonial discourse and the representation of Third World women. The next section deals with the issue of how modernist discourse has pervaded the policies of international development agencies. Then I focus more specifically on the WID regime, as it has been developed by the World Bank, and its limitations. Finally, I discuss the importance of formulating an alternative, Third World women's or empowerment perspective, to overcome the pitfalls of the liberal WID regime. This perspective draws on the insights of both postmodernist and socialist feminism.

26

COLONIAL DISCOURSE AND THE REPRESENTATION OF THIRD WORLD WOMEN

Edward Said's *Orientalism* (1978) was the first systematic study of the cultural production of the "Orient" and the power relations that underlie the scholarly, artistic and government representations of the "East."[2] Said suggests that orientalism constitutes a geo-political/social/cultural distinction and separation, a binary opposition between the manifestation of the Orient and the Occident. It is a powerful device used in various ways to create, reinforce, mystify, manipulate and control the image of the "other," always from the "positional superiority" of the West (Said 1978: 1–31). Although Said's work only makes passing references to gender, his discussion of Flaubert's representation of an Egyptian courtesan, Kuchuk Hanem, is indicative of the colonial discourse on women. Kuchuk Hanem does not speak for herself, but is represented by Flaubert. His description of her ultimately "produced a widely influential model of the Oriental woman" (Said 1979: 6).

Other scholars have explored in greater depth the gendered, male/Western dimensions in colonial/neo-colonial and postcolonial discourse (Alloula 1986; Enloe 1989; Kabbani 1986; Mohanty 1991a; Schick 1990; Spivak 1988). From these accounts it is clear that (neo-)colonial discourse represents Third World women in various ways. The first one is that of the *zenana* representation, whereby veiled Third World women are "mindless members of a harem, preoccupied with petty domestic rivalries rather than with artistic and political affairs of their times" (Enloe 1989: 53).[3] According to this image Third World women are relegated to the *zenana* as housewives, cloistered within the confines of a patriarchal male-dominated environment. In this (re)presentation, Third World women are limited by Third World men to the private sphere dealing with the trivialities of the world within the *zenana*. The public sphere is the preserve only of men who define the structures and the role of the private sphere. In this *zenana* representation the public/private divide is riddled with the double standard of colonial hierarchy. While the public/private divide has been used to relegate both Western and Third World women to the household, Western women are deemed superior whenever the public/private divide and cultural conventions governing non-Western women are different than those affecting their European counterparts. Third World women are monolithically and singularly represented as oblivious to the "real" world, their lives defined and circumscribed by a male-dominated tradition and unquestioningly accepting their confinement. There may be elements of seduction and mystery in this image of ignorance and mindless obedience, but mostly it represents Third World women as inferior to Western women, who do not wear a veil and who have nothing in common with this tradition-bound image. One could argue that the current practice of mail-order brides from Southeast Asia is another example where the *zenana* image is dominant.

In a second representation Third World women are shown as sex objects. This representation is exemplified in Malek Alloula's exposure of the *Colonial Harem* (1986) and Rana Kabbani's *Europe's Myths of the Orient* (1986). Eroticized, unclothed "native" women representing the need to be "civilized" through their contact with the colonizer may appear, at first sight, to be different from their fully clothed counterpart in the *zenana* representation. However, both images define "Third World woman" as inferior and subjugated – the object of sexual desire. Practices of sex tourism in Thailand and Korea and semi-pornographic postcards advertising lush Caribbean beach resorts are examples of the current deployment of this imagery. In both images Third World women are "not allowed to speak" and are "deeply in shadow" (Spivak 1988: 287). Colonization is then justified as a *"mission civilisatrice"* of the "benign and paternalistic" colonizer rather than a "subjugating and exploitative" practice (Schick 1990: 347).

Thirdly, current Western feminist and Western-trained feminist writing often portray Third World women as victims. These feminists base their analysis and their authority to intervene on their "claims to know" the shared and gendered oppression of women. In so doing, they misrepresent the varied interests of "different women by homogenizing the experiences and conditions of Western women across time and culture" (Goetz 1991: 143). Chandra Mohanty's (1991a) analysis criticizes the colonizing aspects of Western feminist writing. According to her, the monolithic and singular portrayal of Third World women as victims of modernization, of an undifferentiated patriarchy and of male domination produce reductive understandings of Third World women's multiple realities.

In all three representations, Third World women are discursively created, separate and distant from the historical, socio-political and lived material realities of their existence. They share the implicit assumption that Third World women are traditional and non-liberated and need to be "civilized" and "developed," i.e. more like Western women. All three representations display elements of the modernist discourse: particularly the binary oppositional categories like Western and Third World, modern and traditional, and liberated and non-liberated women (see chapters 3 and 9).

INTERNATIONAL DEVELOPMENT AGENCIES AND THE LIBERAL DISCOURSE ON MARKETS[4]

In the 1940s Western scholars and policy-makers believed that Third World nations could be "modernized" through infusions of foreign aid, investment and increasing foreign trade. Established in 1944, the World Bank is part of this legacy. The policies of the World Bank have been influenced particularly by the theories of economic modernization which, in general, have updated colonial discourse and reinforced the Northern sense of difference and superiority by utilizing "a series of opposed definitions and contrasting images which reinforced negative concepts of Asian or African societies in terms of comparisons with western Europe, stressing what non-European societies were not" (de Groot 1991: 111). In creating its account of the development process, modernization theory has utilized the traditional–modern dichotomy to suggest that the goal of all traditional societies is to become modern. In a display of simplistic dichotomization, modernization is equated with Westernization, industrialization, and superiority whereas non-modernity is equated with non-Western countries, tradition and inferiority (Hagen 1962; Huntington 1971; Lerner 1964; Sutton 1963).

Economic modernization theories have also utilized as their epistemological premise the superiority and desirability of Western-style modernization and growth before discussing the invisibilities, complementarities and externalities of development. Theorists such as Walter Rostow (1960), W. Arthur Lewis (1966) and P.T. Bauer (1984) maintained that the economic structures of developing countries were riddled with bottlenecks and structural constraints like low savings and consequently low capital stock, and high population pressures (Meier and Seers 1984). Neo-Malthusian explanations were used to explain hunger and poverty in the Third World. Infusions of foreign aid, investment and technological assistance were deemed necessary for growth.[5] Initially, the state was seen as playing a supportive role in this process, although the importance of individual entrepreneurship was never questioned. In the 1980s, the preferred role of the state shifted; it became part of the problem. Comprehensive central planning was seen as an obstacle to growth, diverting scarce resources from development and reinforcing authoritarianism in the Third World (Bauer 1984). The state's function should be to provide the legal and constitutional framework in which individual initiative, private enterprise, foreign assistance, trade and investment can flourish.

From this account it is clear that the liberal discourse on markets permeates the writings of development economists. It is based on the concept of negative liberty developed within the Anglo-Saxon tradition. Individuals are free if the actions of others do not hinder them from their chosen activities and goals (Berlin 1971). Individuals are also the main actors in a (modern) market economy. Capitalist markets, suggest their supporters, are based on voluntary exchanges that bring about a non-coercive, efficient and effective system of production and distribution (Friedman 1962; Nozick 1974; Rothbard 1977). Markets are seen as systems "capable of organizing the cooperation of millions of people" (Lindblom 1977: 9) without coercion (Lindblom 1965: 3), for their mutual benefit (Friedman 1962).

Thus the concepts of unrestrained, freely negotiated, and therefore voluntary exchanges (as performed by modern, male individuals) are seen as central to capitalist markets. Modernization theory is an extension/derivative of the liberal discourse on markets, because it is not only based on the same premises of freedom and individualism, but it is also the goal of modernization theory to turn Third World countries into capitalist market economies (Pieterse 1991).

In the incidental and sparse remarks that modernization theory makes about women, women are essentially represented as tradition-bound conservatives and therefore obstacles to modernization (Jaquette 1982).[6] Similarly, the liberal discourse on markets also reveals a gender-bias. Both are embedded in Enlightenment assumptions, which are imbued with a masculinist (and modernist) epistemology and ontology characterized by a self–other dualism (Eisenstein 1981; Elshtain 1981; Okin 1979; Pateman 1988; Sargent 1981). Other dualisms like "subject–object, mind–body, public–private, fact–value, exchange–use are variations on this theme" (Hirschmann 1992: 228). As Nancy Hartsock (1984) points out, this type of masculinist ontology has given birth to the market model of community based on competition rather than cooperation. The creation of the liberal, capitalist market is then a reassertion of masculinist identity particularly as a desire to dominate others.

The policies of the World Bank have been embedded in the dominant development discourse of modernization. James Ferguson's study of a World Bank project in Lesotho (1990) illustrates how this discourse has disempowered Third World societies and citizens (in particular, women). He exposes the construction of Lesotho as an object of development. According to Ferguson, the development discourse employed by the World Bank first constructed Lesotho as a "solvable" (development) problem. Much like the language of colonial discourse, the World Bank country report classified Lesotho as a "traditional peasant society," a "subsistence economy," which is "virtually untouched by modern economic development" and not yet fully integrated into the regional economy of Southern Africa. Once Lesotho was constructed in these terms, the necessity of World Bank interventions (in the form of funds and expertise) was legitimated. Secondly, the World Bank's development discourse and practices depoliticized Lesotho's "development problems" which were presented as technical and therefore not appropriate for political debate, especially by the poor. This construction contrasts sharply with current academic scholarship which identifies Lesotho as a country with a long history of involvement in the South African political economy and therefore considerable knowledge that could be tapped for development in the region (Ferguson 1985; 1990).

The liberal discourse on markets has defined the World Bank in all its policy phases.[7] However, the implications of this approach have had varying effects on Bank policies towards women. Initially the grounding of its development policies in the liberal discourse on markets contributed to women's invisibility in the Bank. The public/private divide used to separate men and women in the Western world, buttressed by Western patriarchal notions of motherhood and the *zenana* representation of women, contributed to the exclusion of Third World women from early development projects. Early policies assumed that economic development in the public, largely elite male sphere, would naturally "trickle down" to women in the private sphere (World Bank 1990; see early IBRD Reports). The next section will show how this changed in the 1970s.

LIBERAL FEMINISM AND WOMEN IN DEVELOPMENT (WID) IN THE WORLD BANK

In the 1970s Ester Boserup's pioneering work *Women's Role in Economic Development* asserted that modernization had marginalized women and their contributions in the Third World. Her book inspired considerable scholarship on the issue of women's marginalization in development (Arizpe 1977; Buvinic *et al.* 1978; Dixon 1978; Dauber and Cain 1981; Lewis 1984; Staudt 1986; Tinker and Bramsen 1976), along with demands from feminist development groups that women be integrated into the development process (soon known as WID) (Staudt 1985). The UN Decade for Women, the lobbying efforts of women's groups within and outside the World Bank, and the Bank's shift to basic needs under the leadership of McNamara led to the creation, in 1975, of the post of adviser on WID in the Bank. Early efforts included gender training workshops to make reluctant Bank staff more sensitive to women's roles in development, and a report *Recognizing the "Invisible" Woman in Development* (World Bank 1979a).[8] In 1987 the Bank established a Women in Development Division and declared WID an area of "special operational emphasis." Recently, WID coordinators were placed in each of the Bank's four regional complexes (Africa; Asia; Europe, the Middle East and North Africa; and Latin America and the Caribbean) (World Bank 1990: 8–9).

WID projects initiated by the World Bank can be largely classified into three approaches: the welfare approach, the anti-poverty approach and the efficiency approach (Moser 1993).[9] In practice they are often connected. However, in the next section they will be discussed separately to show the different ways in which they intersect with colonial and liberal discourses.

The welfare approach, premised on the assumptions of women as mothers performing childrearing tasks, identified women solely in their reproductive roles. Thus family planning programs, and nutrition projects for children, pregnant and lactating women are examples of a welfare approach. This approach was an expression of early liberal feminist scholarship on development issues and the Bank's top-down approach to development, which was so prominent in the 1960s and early 1970s. It is a reflection of the *zenana* representation of Third World women, pandering to stereotypical images about the roles of Third World women. This approach has never challenged the Bank's dominant discourse on markets, and consequently soon became the preferred approach to WID issues. Indeed, despite heavy criticism, this approach continues to underride many Bank projects. In its own progress report the WID Division of the Bank justifies the continuation of the welfare approach by arguing that "not all operations in all sectors are equally important for actions related to women. Operations in the area of human resources – education and population, health and nutrition are of prime importance" (World Bank 1990: 14; 1993). Compared to projects based on the other approaches, welfare projects receive the highest percentage of funds allocated to the Bank's WID program.[10] Thus "six of the eight projects approved in fiscal 1988 and ten of the eleven in fiscal 1989 do address such basic matters as family planning, nutrition for mothers and children, and maternal and child health care" (World Bank 1990: 15).

The anti-poverty approach emerged in the mid-1970s, partly in response to the criticisms formulated by advocates of the New International Economic Order and dependency theory (see chapter 3). It sought to reduce poverty and provide basic needs for the poor. Women, classified as the poorest of the poor, became one of the targets (Moser 1991). However, although the assumptions of the welfare and the anti-poverty approach appear somewhat different and even conflictual, in practice they have often been similar (Buvinic 1986). Since anti-poverty projects focused on basic needs, they clustered around issues of population and family planning. Indeed, the World Bank's 1979 *World Development Report* addressed women largely in relation to nutrition, family planning and health care (1979b: 83–4, 90). Some anti-poverty projects emphasized income generation, but they concentrated on traditional activities of rural women. They reflected the representation of Third World women discussed above. Educational projects also emphasized the need to educate women because "the influence of the mother's education on family health and family size is great – greater than that of the

32

father's education. Maternal education may also have a greater effect on children's learning" (World Bank 1990: 5).

As with the welfare approach, the anti-poverty approach is embedded in colonial discourse. Not only are "Western frameworks" used to analyze the material realities of women in the South, but Third World women are also portrayed as traditional, voiceless and a homogenous (interchangeable) group. Third World women are presented as the hapless victims of endless pregnancies, bowed down by poor health, illiteracy and poverty (World Bank 1979b). While there are elements of truth in this representation, it ignores the structural causes of poverty. Moreover, it dismisses Third World women's agency and reinforces the myth that Third World women are a homogenous group, without voice or skills. In this approach, Third World women have been objects that need help, not subjects who could be active participants in the development process.

The efficiency approach emerged in the early 1980s and fits readily into the World Bank's turn to neo-classical economics, with its emphasis on structural adjustment programs (SAPs), privatization and liberalization of markets. Since women constitute more than fifty percent of the world's human resources, the efficiency approach urged development efforts to recognize the contributions of women and to integrate them into the development process. This would lead to more efficient growth. It is assumed that issues of equity would be resolved as overall growth occurred.

The current structural adjustment strategies are examples of the efficiency approach. However, structural adjustment has impacted negatively on most poor women, increasing their workloads and the costs of production while decreasing social services and throwing the burden of this reduction largely on women's unpaid labor (Commonwealth Expert Group 1989; Palmer 1991; Staudt 1985; World Bank 1989a;). Even programs introduced to mitigate the social costs of adjustment rely on women's unpaid labor. For example, in Tamil Nadu, India, the World Bank has sponsored a nutrition program for children; low-weight children between the ages of six and thirty-six months from six districts with the lowest calorie intake were provided with direct nutritional supplements for thirty-six months. Community women's volunteer groups provided the cooking, cleaning and serving services necessary for the successful implementation of this program, yet this labor was never factored into the projects successful efficiency rating (Ribe et al. 1990: 29). SAPs, with their focus on women's productive roles, ignore the work of women as homemakers, no doubt, making it easier to ignore the transfer of wage labor to women's unpaid work.

As with the previous two approaches, the efficiency approach is embedded in, and thus reifies, the colonial discourse about Third World women. Indeed, recent World Bank documents often reflect this perspective. For example, a 1990 technical paper on women and nutrition, claims that "Over their reproductive life span, Third World women conceive and nourish with their own bodies 6–8 children" (McGuire and Popkin 1990: 7). It goes on to declare that "During each stage in the life cycle, females have clear-cut biological, economic, and cultural roles" (14), and that women in the South generally lack self-confidence, education and basic skills, even for feeding children (13). This report, which is just one among many, submerges differences among women in the South and reinforces the construction of Third World women as a homogenous, vulnerable group in need of development by Northern experts (see chapters 11 and 12).

THE LIMITATIONS OF WID'S LIBERAL FEMINISM AND OPPOSITIONAL VOICES

Although WID practitioners have challenged the notion that modernization is not a panacea to everyone's ills, they have remained situated squarely within the modernization paradigm (Chowdhry 1993). Calling for the integration of women into development policies and projects, this approach has reinforced many of the assumptions of colonial (discursive) representations of Third World women and the premises of the liberal discourse on markets. The common thread among these three approaches is that Third World women need help to modernize and that the WID projects of the World Bank are the agency of modernization. *Mission civilisatrice* is at work again, only the colonial state has been replaced by WID and the World Bank.

This is not an uncontested arena. Internal dynamics and power struggles within the Bank create spaces from which criticisms can be voiced. For instance, a recent internal report for the Bank specifically challenges the gender-biases in economic models in general and SAPs in particular:

> The objectives of this paper have therefore been broadened to respond to the dual need to bring out the centrality of women's place in the economy (whether or not this is measured or valued), and to place adjustment in the broader framework of social and economic policies that critically define economic opportunities and constraints for men and women.
>
> (World Bank 1993: i)

However, even this report remains tied to the Bank's fundamental approach to development for it concludes by saying that "Finally, there is at root no disagreement with what are identified as major factors essential to growth-oriented adjustment and poverty adjustment" (26).

It calls for the integration of the gender dimension into this approach, not a fundamental rethinking of development itself. This report supports Goetz's argument that:

> the process of modifying women's projects to fit the blueprint of standard development projects has distorted their original objectives. In the end, they represent no threat to the existing power structures and budget allocations within the development establishment.
>
> (Goetz 1991: 135)

Outside critics have been more vocal. Those sympathetic to mainstream development practices have called for more gender-sensitive economic policies, but they do not challenge the growth model of development (Palmer 1991; Tinker 1990). Socialist feminists have been the first to formulate a fundamental critique of the WID approach (see Introduction/Conclusion). Taking the early abstract critiques of capitalism and patriarchy to their logical conclusion, internationally-oriented socialist feminists argue that WID projects are strategic tools of capitalist expansion (Beneria and Sen 1981; Chinchilla 1977; Hartmann 1981; Mies 1982; Young 1981). The main purpose of these projects is to promote capitalism; the development and assistance of Third World women is incidental. However, despite this critical stance, and the involvement of progressive feminist groups in the South, such as DAWN, international feminists have neither challenged the issue of modernity nor one of its expressions, colonial discourse on Third World women (see chapters 2 and 12).

In contrast, postmodernist feminists have formulated a critique of modernity. With their focus on issues of power, difference and gender, postmodernist feminists offer valuable contributions to theorizing about development.[11] Their focus on deconstruction, language and the power of discourse sensitizes us to the power of narration and helps to expose the geo-political relationships of power that underlie the creation of the Third World woman as the "other." Because postmodernist feminism is suspicious of universal claims (to know), it tends to be open to other views and interpretations (Goetz 1991). This has been greeted warmly by Third World feminists who have become critical of Western feminist and WID claims to "know" their problems and solutions.[12] The postmodern feminist focus on difference and their challenge to dualist, universal knowledge-claims has encouraged some scholars to call for new approaches to development. Parpart (1993) has called for development practitioners to listen to the previously silenced and ignored voices of Third World women. In so doing "the goals and aspirations of Third World women would be discovered rather than assumed, and strategies for improving women's lives could be constructed on the basis of actual experiences and aspirations rather than modern fantasies imposed by the West" (454).[13]

Despite its usefulness, postmodernist feminism is not without its contradictions. As a theoretical project that seeks to expose relations of power, abandon universalisms and valorize difference, it is indeed seductive. However, its concern with the "contestability of every claim to knowledge stemming from particular experiences or identities" can result in an "ineffectual position of endless deferral of epistemic responsibility with a deconstructionist degendering of ontology which renders the need and even the possibility of a feminist politics problematic" (Goetz 1991: 144). We need to consider whether postmodernist feminism can offer policy solutions without raising their story to the status of truth (see chapter 9). If all stories are equally valid, which stories should feminist development practitioners adopt?[14] Is the colonial representation of Third World women as valid as the self-representation of Third World women? Furthermore, do we always disallow structural explanations of systemic constraints primarily because they are universalisms?

THIRD WORLD WOMEN'S FEMINISMS/THE EMPOWERMENT APPROACH[15]

Recently, Third World women scholars have been suggesting alternative approaches to the issue of gender and development. One such approach, the Third World Women's (TWW) or the empowerment perspective, is drawn "less from the research of First World women and more from feminist writings and grassroots organizations of Third World women" (Moser 1991: 106). While drawing on many of the insights of the two oppositional perspectives of socialist feminism and postmodernist feminism, the empowerment perspective argues for a development that is more squarely embedded in the particular experiences faced by women and men in the South.

Writings in this perspective initially tended to adopt a political economy approach. Indeed, the DAWN group and its manifesto, *Development, Crises, and Alternative Visions* (1987) influenced socialist feminist thinking on development and contributed greatly to the gender and development approach (Parpart 1993; Rathgeber 1990). The socialist feminists' focus on poverty and the global process of capital accumulation, the complicity of states with the world capitalist system in controlling the productive and reproductive capacities of women, and the intersection of gender with class, contributed to much of the early empowerment writing (AAWORD 1986; Pala 1977; Sen and Grown 1987). This approach, with its stress on political and economic issues, continues to predominate among many Southern feminists. This is particularly true in writings (and activism) concerning structural adjustment, the environment, civil society and the democratization movements (see Chatterjee 1993; Heyzer *et al.* 1995; Nzomo 1992; *Review of Women's Studies* 1991–2; Shiva 1988).

More recently, some Third World scholars and activists have begun to incorporate postmodern concerns, particularly the focus on language, dualist thinking and the construction of the colonial "other" in colonial/ neo-colonial discourses. A number of scholars from the South (although some live and write in the North) are at the forefront of these debates, particularly the focus on postcolonial writings as a source of resistance to colonial/neo-colonial domination (Guha and Spivak 1988; Lazreg 1988; Minh-ha 1987; 1989; Mohanty 1991a; Prakash 1990; Spivak 1987; 1988).[16] The esoteric language and literary focus of much of this writing has impeded its adoption by many Third World activists, but the key elements of this approach, particularly the focus on language and representation, are increasingly influential in empowerment writings (Barriteau Foster 1992; Heyzer and Wee forthcoming; Vargas 1992).

Whatever their intellectual roots, scholars working with these new ideas are deeply suspicious of existing modernization projects, which they see as a cause of maldevelopment in the Third World (Sen and Grown 1987; Shiva 1988; Steady 1987). Since the modernization project protects and expands the interests of capitalism and the Western world, it is viewed as fundamentally in conflict with the needs and goals of Third World women. From the empowerment perspective, modernization and its intellectual production is complicit with Western interests. Some scholars regard socialist "solutions" in the same light (Spivak 1988: 2; chapter 8).

Rooted in the concrete and contextual realities, experiences and wisdom of Third World women, this emerging body of literature (and activism) calls for a new kind of thinking and action. While drawing on Western scholarship, scholars and activists in this perspective are determined to develop an approach that is rooted in the experiences of women (and men) in the South rather than those of Western women. They call for a renewed focus on indigenous grassroots movements. As Shiva points out, these movements:

> are creating a feminist ideology that transcends gender and a political practice that is humanly inclusive; they are challenging patriarchy's ideological claim to universalism not with another universalizing tendency, but with diversity; and they are challenging the dominant concept of power as violence with the alternative concept of non-violence as power.
>
> (Shiva 1988: xviii)

The approach of WID projects which portray women as victims and men as the beneficiaries of modernization, and then seek to right these wrongs of modernization, is discarded by the empowerment perspective. Instead, women and men are not necessarily poised antagonistically against each other, nor are all women joined by the invisible strands of sisterhood. Elements of class, ethnicity and race intersect with gender to form alliances between men and women. This perspective focuses on grassroots organizations that seek to empower women by increasing their collective capacity towards self-reliance. They call for the redistribution of power, both intranationally and internationally, so that poor women can participate in controlling and influencing the directions in which development occurs (Antrobus 1989; Asian Women Writers' Workshop 1988; Chhachhi 1988; Moser 1993; Sen and Grown 1987).

The empowerment approach has had little influence on mainstream development agencies, although a few more progressive agencies have begun to adopt a Gender and Development approach which at least discusses the need for addressing gender subordination in the South (CIDA 1992). However, it has influenced and been influenced by grassroots organizations around the world.[17] Groups like Gabriela in the Philippines, the Self-employed Women's Association (SEWA) in India, and the Grameen Bank in Bangladesh are examples of the success of the strategy of empowerment through organization (Moser 1991; 1993; see chapter 8).

The adoption of empowerment strategies by grassroots organizations is a practical alternative to the top-down strategies adopted by WID. It reminds development agents that development, particularly of women, cannot ignore the specificities of life in the Third World. However, it does have a tendency to lapse into a romantic, essentializing vision of Third World women (see chapter 2). For instance, in their effort to valorize poor Third World women, some Third World ecofeminists have overemphasized the nurturing role of Third World women (Mies and Shiva 1993). While politically strategic, this construction comes dangerously close to the essentialist assumptions and strategies of WID.

CONCLUSION

The liberal feminist struggle to establish WID has at once succeeded and failed. It has succeeded in establishing WID divisions within international development regimes and fostering mostly separate projects for women. Its epistemological foundations and existence within these development agencies, however, have limited it from moving beyond the assumptions of modernization, of whose impact it was critical.

What are the lessons development agencies can learn from the varied oppositional approaches to gender and development? Are the criticisms addressed at WID, suggesting *recognition* of the diversity of Third World women's roles rather than their *integration*, merely the semantic prefer-ences of armchair academics with no implications for the implementation of the development process? Will it make a difference if developmentalists recognize the diversity of Third World women's roles, skills, experiences, needs and aspirations? Will it make a difference if we listen to the previously silenced voices of Third World women?

The lessons of the oppositional approaches to WID are many. The socialist feminists' focus on gender and class, combined with the postmodernist feminists' focus on difference and the power of narration, provide valuable insights for development scholars and practitioners. The empowerment approach incorporates the strengths, but not the weaknesses of these approaches in its analysis. It addresses both the theoretical and practical needs of the "crisis" in development. Theoretically, Third World affirmative, postmodernist feminists provide us with the tools to deconstruct the discourse of development and expose its Western bias. They affirm the multiple realities of women, particularly their situated, localized character. Like socialist feminists they focus on the material realities of Third World women within the world system. Like postmodernist feminists they add other differences like ethnicity, race and age to the socialist feminist difference of class and gender. Amidst this theoretical deconstruction, they ask us to listen to Third World women's voices, to reconstruct their realities within the intranational and international systems of power. To them all stories do not carry the same weight. They affirm the stories of Third World women in which Third World women speak for themselves or other authors speak "near-by" or "together-with" rather than "for and about" Third World women (Minh-ha 1987). In doing so, Third World women become participants in, rather than recipients, of the development process.

NOTES

1 I thank the editors for their extensive revisions. An earlier version was presented at the International Studies Association, Vancouver, BC, Canada, March 1991.

2 The work of Albert Memmi, *The Colonizer and the Colonized* (1967), is also a powerful narration of the successful creation of the other as object as well as the recreation of the self as subject.

3 *Zenana*, in Urdu, are the private quarters of women, inside which men rarely entered. On the rare occasions that women did venture outside, they were always in purdah (veiled).

4 Of course, not all development agencies are the same. They often have particular emphases and are constrained in various ways by their constitutional mandates. However, since they are the product of the same historical and paradigmatic era, they generally adhere to the same paradigm of development. Hence there are broad similarities among most, particularly mainstream, development agencies.

5 P.T. Bauer believed in some cases foreign aid actually supported recalcitrant

regimes and thus inhibited development (Bauer 1984: 42). According to Rostow, Bauer believed that foreign aid should be used to promote private enterprise and Western style democracy (Rostow 1984: 246).

6 Jaquette notes that Talcott Parsons is one of the few modernization theorists who has anything to say about women in the Third World. He calls them tradition-conservationists (Jaquette 1982). It is ironic that some Third World feminists use the same description of Third World women to say how important women would be in a development that seeks to conserve culture (Jayawardena 1986).

7 The World Bank's policy phases have ranged from trickle-down to redistribution with growth. Although redistribution with growth, with its focus on an increased role of the state, evoked criticism from the conservatives, it lay squarely within the domain of the liberal discourse on markets. State power was strengthened in order to guarantee the proper functioning of markets and the success of capitalism (Payer 1974; 1982; Chowdhry forthcoming).

8 The directors of a training workshop on Women in Agricultural Development (WIAD) at the University of Florida regaled us with tales about the hostility and resistance of many Bank staff to gender training.

9 Buvinic (1986) used the categories of welfare, equity and anti-poverty to classify policy approaches to women. Moser (1991; 1993) adds efficiency and empowerment to those categories. It should be noted that these approaches are academic constructs which artificially classify projects into categories. They often overlap and projects often have elements of all of them.

10 Projects that specifically address women increased from 9 percent in 1980–7 to 30 percent in 1988–9 in agriculture, from 22 percent to 33 percent in education and remained constant at 75 percent for population, health and nutrition.

11 I participated in a roundtable on feminist postmodernism and development at the 1991 conference of the Association for Women in Development (AWID). The audience included scholars, advocates and practitioners. The session evoked some very emotional and hostile comments from practitioners questioning the usefulness of this approach for them.

12 For a critical look at Third World women's outrage at the practices of Western feminism see Nawal El Saadawi, Fatima Mernissi, and Mallica Vajarathon (1978); Sen and Grown (1987).

13 I am not merely pointing to foreign consultants, but also to Third World experts, who are being used, albeit sparingly, as consultants as well, some of whom see their clients as the underdeveloped "other."

14 I am indebted to Goetz (1991) for this idea.

15 It may seem ironic to use the homogenizing category of Third World women's perspectives after using that as a critique of other perspectives. However, I deliberately use the plural when discussing these perspectives to suggest a diversity of thinking ranging from greater leanings towards socialism, ecology or postmodernism. The term has been used most prominently by Moser (1991; 1993), but she offers a very open, fluid "definition" of the approach. It awaits further clarification and definition.

16 Pauline Rosenau (1992) has classified them as Third World affirmative postmodernists. I prefer to classify them in the Third World feminist/empowerment perspective. Although they use postmodern tools of deconstruction, they have also taken issue with postmodernism. Also, often they

have classified themselves as Third World feminists or Third World women rather than feminist postmodernists.

17 Countries like Canada and Norway have been more supportive of the empowerment approach and have provided some funds to these groups (Boesveld *et al.* 1986).

2

WOMEN AND DEVELOPMENT

A critique

Mitu Hirshman[1]

This chapter engages in a critical discussion of Gita Sen and Caren Grown's much-acclaimed monograph, *Development, Crises, and Alternative Visions*.[2] In addition to providing a detailed discussion of the development process in Third World countries and a poignant report on the "less than benign" development experiences of poor Third World women in the rural areas (note: not all women across the board), the authors suggest some possible remedies in the form of "alternative blueprints and strategies" to current development practices. These "alternative visions," which include a potpourri of items ranging from tree planting for fuel needs in a local village in South Asia to peace justice and equality on earth for all, purport at the same time to give the Women in Development (WID) enterprise a much-needed reorientation and shift in focus. For Sen and Grown, it is not enough (though necessary) for WID to point out certain inherent sex and gender-related biases in mainstream development paradigms; it is necessary to establish an integral link between effective development and the eradication of the social oppression and material poverty of "poor women." Development theorizing as well as planning should begin from the vantage point of "poor women" because:

> The perspective of poor and oppressed women provides a unique and powerful vantage point from which we can examine the effects of development programmes and strategies. . . . The vantage point of poor women thus enables us not only to evaluate the extent to which development strategies benefit or harm the poorest and most oppressed sections of the people, but also to judge their impact on a range of sectors and activities crucial to socio-economic development and human welfare.
>
> (Sen and Grown 1987: 23–4)

But why "poor women"? Following the logic of *feminist standpoint theory*[3] the authors argue that if the objectives of the development enterprise:

42

include improved standards of living, removal of poverty, access to dignified employment, and reduction in societal inequality, then it is *quite natural* (italics mine) to start with women. They constitute the majority of the poor, the underemployed, and the economically and socially disadvantaged in most societies. Furthermore, women suffer from the additional burdens imposed by gender-based hierarchies and subordination. *Second*, women's work, under-remunerated and undervalued as it is, is vital to the survival and ongoing reproduction of human beings in all societies.

(Sen and Grown 1987: 23–4)

As Sen and Grown point out, poor women perform a disproportionate amount of the basic work required to sustain societies in the South, particularly reproductive work. Moreover, this burden has intensified as the economies in many developing countries have struggled with economic decline and the harsh structural adjustment packages prescribed by the World Bank and International Monetary Fund (IMF). The authors conclude that solutions to given developmental problems should focus on women, whose experiences and viewpoints "as the principal producers and workers is an obvious starting point" (Sen and Grown 1987: 24).

In other words, according to Sen and Grown, provision of minimum basic needs for all should be the underlying guiding principle of development planning and policy-making. Since women constitute the human element linking the availability of food, rural energy sources and water in many parts of the world, a sound and effective development practice should, therefore, place "poor women" at its center (Sen and Grown 1987: 57–8). Ignoring this *essential fact* has led to the failure of development practice (grounded, as the authors suggest, in a distinctively male ethic). This approach has also failed to relieve overall poverty and social oppression. Indeed, it has helped to worsen the already deteriorating conditions surrounding women's existence.

Sen and Grown, in choosing to examine and assess the development enterprise from the vantage point of "poor women," follow Marx, who examined and analyzed the capitalist mode of production from the vantage point of the proletariat or working class. Like the lives of proletarians, they believe poor women's lives and experiences in development offer "a particular and privileged vantage point" that can provide the grounding for a powerful critique of development, revealing it as an offspring of phallocentric institutions and ideology:

it is from the perspective of the most oppressed – i.e., women who suffer on account of class, race and nationality – that we can most

43

clearly grasp the nature of the links in the chain of oppression and explore the kinds of actions that we must take.

(Sen and Grown 1987: 20)

And so,

it is the experiences lived by poor women throughout the Third World in their struggles to ensure the basic survival of their families and themselves that provide the clearest lens for an understanding of development processes. And it is their aspirations and struggles for a future free of the multiple oppressions of gender, class, race, and nation that can form the basis for the new visions and strategies that the new world now needs.

(Sen and Grown 1987: 9–10)

In keeping with the Marxian logic, therefore, socio-economic and political freedom for all ("poor" and "women") is predicated upon the complete emancipation of "poor women," victims of race, class and gender oppression. Furthermore, like the proletariat's indispensable role in the functioning of the capitalist machinery, "women's contributions – as workers and managers of human welfare – are central to the ability of households, communities, and nations to tackle the current crises of survival" (Sen and Grown 1987: 18). In other words, women's labor and efforts support the vital sectors of society, namely, food production and processing, child bearing and rearing, nurture and care without which the continued survival of society (in Third World countries) becomes impossible. Yet, development practitioners, the authors claim, tend to leave out this crucial fact of women's participation in societal production and reproduction and with sheer "ideological blindness" pursue policies that worsen the already oppressive and exploitative conditions originating in gender-based ideologies and discriminations surrounding women's existence.

The poignancy with which Sen and Grown (and in this they follow their predecessors in the WID enterprise) report on the oppressive conditions of Third World women makes it difficult to launch a critique, let alone political opposition to their position. To call into question the assumptions and practices of this approach is to bring down upon our heads the opprobrium of many different, and even incompatible, schools of thought informing development theory and practice. It is to turn one's back on the most noble of activities, the be-all and the end-all of the humanist project: the improvement of the human condition. My object here is not to cast doubts on the sincerity of the WID practitioners in their quest for a "true" development that will benefit one and all. My purpose in this chapter is to argue that despite the authors' claim that DAWN represents a radical alternative to both conventional development theory and practice, as well as to WID, it shares with them certain basic premises and assumptions about both "development" and "women." In this context, I would like to draw attention to three critical issues related to Sen and Grown's discourse on "alternative development" issues that have been in the forefront of current feminist debates. One rather controversial issue concerns the deployment of the category, "the sexual division of labor" or more generally the concept of "labor" for understanding women's oppression and social subordination. Another issue focuses on the relationship between the sexual division of labor and the construction of emancipatory forms of knowledge. Lastly, much has been written about the representation of Third World women as *victims* in the narratives of both Western feminism and WID practitioners. I shall discuss each of these in turn.

THE SEXUAL DIVISION OF LABOR

In giving analytic primacy to the concept of women's labor, or more generally, the sexual division of labor, in their theory of alternative development, Sen and Grown have made themselves vulnerable to charges of essentialism, foundationalism and ethnocentrism.[4] By positing "poor women's labor" as the defining category and the founding source of women's experiences in the South, and also as the grounds for their alternative approach to development, the authors commit themselves to a form of essentialism which seeks to establish a priori an indisputable natural and innate essence to Third World women's lives and experiences. This is derived not necessarily from "biological facts," but from secondary sociological and anthropological universals, which define the sexual division of labor. Furthermore, they postulate in advance the kind of experiences deemed significant in understanding women's lives, thereby marginalizing and repressing "other" experiences that cannot be explained or accounted for within this theoretical logic. In short, the "sexual division of labor" and "women" are implicitly treated as "commensurate analytical categories" outside of race, class, history and culture.

As Chandra Mohanty points out, this is a common practice. In the search for explanations to women's subordination, many feminists (particularly in the North) have sought explanations in the universal applicability of concepts such as reproduction, the sexual division of labor and patriarchy. Mohanty questions this approach. How, she asks:

> is it possible to refer to "the" sexual division of labour when the *content* of this division changes radically from one environment to the next, and from one historical juncture to another? At its most abstract level . . . concepts such as the sexual division of labour

45

can be useful only if they are generated through local, contextual analysis. If such concepts are assumed to be universally applicable, the resultant homogenization of class, race, religious, and daily material practices of women in the third world can create a false sense of the commonality of oppressions, interests, and struggles between and among women globally.

(Mohanty 1991a: 67–8)

Sen and Grown, by privileging the sexual division of labor and treating it as an unquestionable, permanent and transcendent category give "a false sense of legitimacy and universality to a culturally specific, and in some contexts culturally oppressive" definition of "woman" and "woman's work". By making women's labor (both as producers of basic needs and reproducers of human beings) the structural determinant to women's existence, they reduce the multifarious reality of "women's being" to this single logic of production and labor.[5]

This is not to deny, however, that women do have some experiences in common underlying their differences. But the thorny issue that has plagued the international women's movement, and the WID approach in particular, is the question: How can we effect a reconciliation between the presumed universal nature of gender experience (i.e. oppression) and individual women's concrete experiences embedded in their member-ships in many different and often conflicting social and political groups (class, caste, political parties, race, ethnic and nation-state, to name a few) as well as with diverse cultural practices and symbolic forms. A related question, of course, is the extent to which global sisterhood is a viable political project, given the diverse experiences of women's lives. As Mary Poovey cautions:

> We need to remember that there are concrete historical women whose differences reveal the inadequacy of this unified category in the present and the past. The multiple positions real women occupy – the positions dictated by race, for example, or by class or sexual preference – should alert us to the inadequacy of binary logic and unitary selves without making us forget that this logic has dictated (and still does) some aspects of women's social treatment.
>
> (Poovey 1988: 62–3)

Sen and Grown too caution the WID enterprise against adopting a concept of gender that ignores differences and diversities.

> We strongly support the position in this debate that feminism cannot be monolithic in its issues, goals, and strategies, since it constitutes the political expression of the concerns and interests of women from different regions, classes, nationalities, and ethnic backgrounds. While gender subordination has universal elements, feminism can-

not be based on a rigid concept of universality that negates the wide variation in women's experience . . . feminism cannot be monolithic in its issues, goals and strategies, since it constitutes the political expressions of the concerns of women from different regions, classes, nationalities, and ethnic backgrounds. There is and must be a diversity of feminisms, responsive to the different needs and concerns of different women, and defined by them for themselves. This diversity builds on a common opposition to gender oppression and hierarchy.

(Sen and Grown 1987: 18–19)

Yet for all their awareness of "diversity," Sen and Grown believe this diversity "builds on the common opposition to the immutable 'fact' of gender oppression and hierarchy." Thus they fall into the trap of "universalism." In the final analysis, such an acknowledgment of the diversities of women's "experiences" of oppression (read: benevolent pluralism) "does not change the nature of things." It simply preserves the "Woman" of Western feminism (like the "Man" of Western human-ism).[6]

In the end, however, privileging the "woman question" (pitted against "man") appears to be a political necessity for Sen and Grown and their feminist project. They argue that differences should be set aside as less important than that which connects women, because only then will women be able to unite against men and male oppression. Difference is seen as an impediment to alliances between women of various denominations and diverse locations in terms of race, class and ethni-city, and nationality. They adopt this position in the belief that:

this analysis can contribute to the ongoing debate about the com-monalities and differences in the oppression of women of different nations, classes, or ethnic groups. "Sisterhood" is not an abstract principle; it is a concrete goal that must be achieved through a process of debate and action.

(Sen and Grown 1987: 24)

They believe it is "At the global level, [that] a movement of women and the oppressed can mobilize support for the common goals of a more just and equitable international order" (Sen and Grown 1987: 87).

This attempt to resolve the differences among women with a single stroke – i.e. to choose "global sisterhood" for political reasons rather than because it is a genuine possibility – has been read by some Third World feminists as an attempt on the part of Western feminism to colonize and domesticate the multifarious realities of "being women" in non-Western societies under the sign of a singular, homogenous Third World Woman (Lazreg 1988; Mohanty 1991a; Spivak 1988). "Global sisterhood," it seems to me, is postulated as the ultimate goal and invoked frequently during the course of their discussion as the panacea to the fragmentariness engendered by the fact that individual women occupy different and often conflicting positions intersected by race, class and ethnicity. Echoing Hartsock's confession,[7] Sen and Grown bracket "differences" among women by invoking the common background to women's oppression and subordination: male control of resources and power, and the divisions of labor that have enshrined male privileges.

The "sexual division of labor" is thus taken to be "a universal structure" that constitutes the basis for an "essential woman" oppressed by male domination. This system of subordination provides the basis for building a different and better society: *a better alternative*. But what this discourse of an alternative development fails to recognize is that concepts such as "labor" and "production," which are integral to the theory of "the sexual division of labor," have been challenged by postmodernists such as Baudrillard, as well as by feminists (Flax (1987), Butler and Scott (1992) in particular). These critiques question representations which ignore and indeed "other" the lives of Third World peoples, especially women. The concepts of "labor" and "production," as Baudrillard maintains, are categories too rooted in the culture of capitalism and modernity and are therefore inadequate when it comes to describing "other" societies – be they preindustrial, feudal, "archaic" as well as many contemporary non-European and non-capitalist societies and must therefore be abandoned. "All these concepts" says Baudrillard, "are in fact historical products" which have acquired the status of cultural universals through questionable power moves. Implicit in the concepts of "labor" and "production" is a "productivist ideology" which views human beings as laboring machines "seeking [their] telos in the conquest of nature" (as well as women and "others") in order to fulfil the capitalist goal of "unlimited and unchained productivity." In fact, an uncritical deployment of these concepts in order to comprehend the social realities of "other" people not only distorts life in other cultures but "serve[s] to subordinate and erase that which they seek to explain." As Baudrillard asks:

Does the capitalist economy retrospectively illuminate medieval, ancient, and primitive societies? No: starting with the economic and production as the determinant instance, other types of organization are illuminated only in terms of this model and not in their specificity or even, as we have seen in the case of primitive societies, *in their irreducibility to production*. The magical, the religious, and the symbolic are relegated to the margins of the economy.

(Baudrillard 1975: 86–7)

48

The concepts of labor and production, also, are theoretical categories abstracted from the historical experience of a specific group – the European Man. They hypostatize the culture of Enlightenment and put at the center "a very European Man" whose historical emergence on the world stage intensified the repression of not just women but also of "other" people (Flax 1990b; Harding 1991; Young 1990). Marx too shares in this Enlightenment ideology as evidenced in his conception of labor, or more specifically, productive labor, which is defined as the means by which man reproduces himself as man in his own likeness. Marx thus equates human beings with being male. Also, the concept of productive labor is modeled on traditional masculine activities in modern, Western society, particularly production. Thus, Marx's categories despite their radical promise, are not free from the gender and race bias of bourgeois philosophy. The Marxian imaginary simply reproduces the androcentric and ethnocentric[8] bias inherent in the Enlightenment philosophy. By privileging the productivist paradigm, which is itself a product of capitalism, Marxists foreclose the possibilities for a more radical alternative.

So the question is: Should feminists borrow concepts without considering the cultural and historical limits of these concepts? For instance, socialist feminists' attempt to widen the Marxist paradigm so as to overcome the gender bias of Marxist theory by expanding the concepts of labor and production to include many of the activities associated primarily with women's work (such as the production of children and reproduction of human beings). But they have not overcome the limitations of the Marxist framework itself. Instead:

This explanatory framework incorporates the historical and philosophical flaws of Marxist analysis. Marxists (including socialist feminists) uncritically apply the concepts Marx should have used to describe a particular form of the production of commodities to all areas of human life at all historical periods. These concepts include the categories of labour (e.g., productive or nonproductive), the centrality of production in the organization and culture of any society, the importance of exchange, the creation of surplus value and commodity production within the "economy," and the definition of class solely in terms of the *individual's* relation to the means of production. Marx and subsequent Marxists replicate rather than deconstruct the capitalist mentality by essentializing what is in fact a product of a particular historical and variable set of social relations. Such theoretical moves not only distort life in capitalist society but surely are not appropriate all cultures.

(Flax 1990b: 154–5)

In the last analysis, the categories of labor and production, as conceptualized in Marxist theory, are no more free from gender and race bias than those of their bourgeois predecessors. It will not do to simply revise or extend those concepts which repress, distort and obscure many aspects of women's existence. Instead of seeking to "widen" the concept of production, why not dislodge it or any other central concept from the "authoritative powers" of Western male discourse?

THIRD WORLD WOMEN AND DEVELOPMENT

Besides poor and disadvantaged Third World women – the victims – the participants in Sen and Grown's development drama include a network of activists, researchers and policy-makers. These committed women and occasionally men, from a number of different countries in the South and North, came together under the auspices of a number of progressive mainstream development agencies such as the Ford Foundation and the Norwegian Agency for International Development (NORAD), to share their experiences of development strategies, policies, theories and research. Like the good Samaritans (reminiscent of the missionary zeal of the eighteenth- and nineteenth-century colonial administrators),[9] many of these development specialists have sought to represent and rescue at the same time the "silent and mute victims" (i.e. poor Third World women)[10] from their own men (symbolizing patriarchal oppression and domination), who are in league with capitalism (symbolizing the hierarchical, oppressive and the exploitative international division of labor), the "bad guys" in other words. Through their analysis of the development process, conjoined with their understanding of feminism, they (the Subject) have sought to empower the oppressed and poor women (the subject/object)[11] in various Third World countries and to engage them in grass roots struggles to recuperate the subjecthood (agency) which they presumably lost, in some cases, during the capitalist penetration of their societies.

While acknowledging that "Women drawing on their experiences have developed great capacities for internal resilience and resistance," indeed they have sought to shape the destiny of their societies (in the process of reshaping their own) through collective nonviolent resistances to issues and forces such as "nuclear weapons, military death squads, and forest contractors," the Sen and Grown book points out that these collective actions have not been particularly transformatory. They have been "small and fragmented" to use the authors' own words (1987: 78). Nevertheless these actions underscore the fact that women in the Third World, if properly mobilized (note: guided and instructed), that is, given voice to, could break out of their "traditional submissiveness" (note: into

50

"modernity") and work towards improving economic conditions for themselves and others.

> Women have learned to shed *traditional submissiveness* (emphasis added) and withstand family and community pressures, and have begun to work together to improve economic conditions for themselves and others. Women have organized to use cultural forms to raise the consciousness of men and women about injustice and inequality.
>
> (Sen and Grown 1987: 78)

Implicit in statements such as the above is the need to show that Third World women as "women" (i.e. victims of men and capitalism) share the same cultural space and political rationality as Western feminism and that they can be relied upon to participate in the feminist politics appropriate to the pursuit of their interests, namely, women's empowerment *vis-à-vis* men and male establishments. Implicit also is "the desire to show that secular, rational politics did not have to be imposed from above. Instead, left to their own invention poor women were capable of great creativity" as exemplified in the women's empowerment movement.

> Indeed with or without international or governmental recognition, women have already been organizing themselves. In India, for example, there is a spreading movement of women organizing against forest contractors to prevent deforestation in India. Traditional kin and community-based system of mutual aid and self-help in Kenya, are vitalized for collective solutions to fuel and water problems.
>
> (Sen and Grown 1987: 58)

This representation assumes:

> an identical cross-cultural universal subordination [and therefore of identical interests and targets. As Young points out, it] also privileges unquestioningly the values of Western feminism, while remaining unselfconscious about its own relation to the oppressive political–economic power structures that operate between the West and non-Western countries. Western feminist discourse, in short, can not only be ethnocentric, but in certain contexts can itself be shown to be a contemporary form of colonial discourse.
>
> (Young 1990: 162)

Just as the colonized have been constructed according to the terms of the colonizer's own self-image, so, in Sen and Grown's book, poor Third World women are also constructed in terms of feminism's own "narcissistic" self-image. It is entirely possible that such a construction may have little or no existence or reality outside its representation. It is time that we rethink and reconstitute feminisms' as well as WID's relationship to "poor Third World women," not by casting them as victims who need help, not by asking "What can we do for them?," thus implying that they have the privileged insight into as well as the only true story about their conditions.[12] Instead of patronizing and romanticizing them, we:

> must learn to learn from them, to speak to them, to suspect that their access to the political and sexual scene is not merely to be corrected by our superior theory and enlightened compassion . . . in order to learn enough about Third World women and to develop a different readership, the immense heterogeneity of the field must be appreciated, and the First World feminist must learn to stop feeling privileged as a woman.
>
> (Spivak 1988: 135–6)

CONCLUSION

To conclude, Sen and Grown have failed to provide an acceptable alternative framework for the development enterprise. Although WID interventions have been a necessary step in the right direction, this approach has proven incapable of challenging gender stereotypes and male structures of power. Sen and Grown have suggested a much needed alternative. But in privileging the "production–reproduction" grid to illuminate the lives of Third World women as well as explain their conditions of subordination and oppression, the women and development approach imports into its discourse and practice the structures and presuppositions of the very systems it opposes – namely, capitalism and patriarchy. By accepting "women's labor" as an unproblematic sociological category and claiming that women's labor has simply been ignored or neglected in conventional development theory and practice, without asking how such neglect has occurred or whether such neglect is integral to the androcentric practice of development, Sen and Grown, like their WID counterparts, fail to challenge the basic premises underlying traditional Marxist (read: androcentric and Eurocentric) thinking, namely that "labor" is the essence of "being human."

In their crusade to build a "broad-based" women's movement at the national and international level so as to bring about "progressive" social change – cutting across not just different women's organizations and political affiliations but also class, ethnic and racial barriers – Sen and Grown blithely ignore and erase all the complexities and concreteness surrounding women's lives, in the same way that "traditional development thinking," in its efforts to push "modernization" and "capitalism," ignored the cultural and historical specificity of various Third World societies. In addition, their analysis suffers from the same economistic bias as mainstream development theory, which is entrenched in the belief that material needs constitute the sole determinant of human existence. Thus it appears that for those practitioners adopting Sen and Grown's approach, the provision of food–fuel–water (reproduction) form the cornerstone of women's existence, bereft of specific histories, cultures and social setting within which such "needs" are articulated. The emphasis, unwaveringly, is on the economic realm of the women's existence. They naively assume that once the bread-and-butter (basic needs) issues are taken care of, other needs of a non-economic nature will fall into place.

While Sen and Grown point to the contextuality of women's lives from time to time, this does not detract attention from the fact that the analytical framework they have adopted to investigate the development experiences of poor Third World women is the production–reproduction grid. This approach, as Lazreg points out "precludes any understanding of third world women in their lived reality and persons in their own right. Instead they are reified and made into the bearers of pre-given categories" (Lazreg 1988: 94). In this way Sen and Grown tend to reduce both the complexity of the development process and women's existence by reducing it to the universal category of either *labor*, such as in the procurement of food–fuel–water, or *gender oppression*, symbolized by exclusion, clitoridectomy, restricted mobility, sexual violence and so forth. Consequently, Sen and Grown's "alternative visions" remain mired in androcentric Western thinking and fail to provide a genuine alternative to mainstream development theory and practice.

NOTES

1 This is a revision of a paper originally presented at the International Studies Association meetings held in Atlanta, Georgia, March 1991. The author thanks Kimberley Flynn, Bettina Zolkower and Roslyn Bologh for their involvement at the earlier stages in the writing of this paper. For comments and criticism I would like to thank Carmen Medeiros, Kriemild Saunders, Yvonne Lasalle, Maureen O'Dougherty and other participants in the *Feminism, Postmodernism and the Third World* study group held at the Graduate School, City University of New York. Finally, my gratitude to Stanley Aronowitz for his continuing encouragement and support in keeping this project alive. Also, my thanks to the editors of the present volume for their comments and criticism which helped to sharpen its argument further.

2 The Gita Sen and Caren Grown monograph presents a theoretical articulation of the DAWN project. The term women and development is used here to describe the alternative visions presented in the document and associated development practices, not the more specific definition of WAD described in the introduction.

3 See Hartsock 1987: 157–80.

4   It is not my purpose to rehearse here the rather large literature dealing with
    these issues variously gathered under the rubric of feminism and postmo-
    dernism, poststructuralist feminism and deconstruction. The interested
    reader can be well guided in this respect by the Bibliography.

5   "No one . . . denies there are universal facts, such as birth or death. But take
    away their historical and cultural context, and anything which is said about
    them can only be tautological . . . the suggestion that work is as natural as
    birth or death negates its historicity, its different conditions, modes, and
    ends – specificities which matter . . . " (Young 1990: 123).

6   "Roland Barthes' short essay, 'The Great Family of Man' . . . [discusses] the
    well-known photographic exhibition which is organized around the fiction
    of the universality of fundamental human experiences . . . .'The Family of
    Man' projects the myth . . . that underneath there is one human nature and
    therefore a common human essence. Barthes argues that such humanism, so
    reassuring at the sentimental level, functions simply to override differences
    'which we shall here quite simply call 'injustices'" (Young 1990: 122).

7   "In addressing the institutional sexual division of labour, I propose to lay
    aside the important differences among women and instead to search for
    central commonalities across race and class boundaries" (Hartsock 1987:
    232). In other words, Hartsock takes for granted that "there are some things
    common to all women's lives in both Western and Third World societies.
    The basis of this commonality is the sexual division of labour [according to
    which] women as a sex are institutionally responsible for producing both
    goods and human beings and all women are forced to become the kinds of
    persons who can do both" (Ferguson 1991: 332).

8   Marx's writings on India is a case in point. "Marx is at his most savagely
    modernist," says Derek Sayer, "in his writing on 'Asia'. Orientalist cliches
    thoroughly permeate Marx's discourse. . . . Two infamous articles of 1853
    are illustrative of his sentiments. 'Sickening as it must be to human feeling,'
    he tells the readers of the *New York Daily Tribune,* 'to witness the disintegration
    of India's ancient village life at the hands of British capitalism, we must not
    forget that these idyllic village-communities, inoffensive as they may appear,
    had always been the solid foundation of oriental despotism, that they
    restrained the human mind within the smallest possible compass, making
    it the unresisting tool of superstition, enslaving it beneath the traditional
    rules, depriving it of all grandeur and historical energies'" (Sayer 1991: 14–
    15). Derek Sayer notes that in "Marx's imagery of Asiatic passivity we catch
    an echo of a common sexualization of 'the West' and its dark and mysterious
    Other, which exists to be possessed. . . . 'Asia' is passively feminine, an
    object of conquest and desire, modernity thrusting, masculine erect"
    (Sayer 1991: 15).

9   "Sympathy for the oppressed is as old as colonialism itself" (Young 1990).

10 Sen and Grown consistently represent women in the South as *victims*: victims
    of the "villainous acts" of their men (sex/gender oppression) and the global
    economic and political system (class oppression). As Mohanty remarks:
    "Although it is true that the potential of male violence against women
    circumscribes and elucidates their social position to a certain extent,
    defining women as archetypal victims freezes them into 'objects-who-de-
    fend-themselves,' men into 'subjects-who-perpetrate-violence,' and (every)
    society [divided] into powerless (read: women) and powerful (read: men)
    groups of people" (Mohanty 1991a: 58).
    Jane Flax, too, points out the pitfalls of an essentialist construction of

women as *victims* and men as *victimizers*. "We need to avoid seeing women as totally innocent, acted upon beings. We should not fall into the victim's viewpoint. Such a view prevents us from seeing the areas of life in which women have had an effect, are not totally determined by the will of the other and the ways in which some women have and do exert power over others (e.g. the differential privileges of race, class and sexual preference, age and location in the world-system)" (Flax 1990: 181–2).

11 "The individual [in this case Third World women] becomes a subject (small 's') by being subjected to the Subject (capital 'S') . . . become[s] reflections of the Absolute Subject (first world feminist). Althusser who uses this idea of the Absolute Subject (referred to as the God in the history of Western thought) as a metaphor to explain how individuals internalize or rather becomes reflections of the ruling ideology, 'reflecting' the ideas and norms of the institutions they find themselves in. These individuals become part of the Absolute Subject" (Assiter 1990: 126).

12 " . . . we grieve for our Third World sisters; we grieve and rejoice that they must lose themselves and become as much like us as possible in order to be 'free'; we congratulate ourselves on our specialists' knowledge of them" (Spivak 1988: 179).

3

LATIN AMERICAN WOMEN SPEAK ON DEVELOPMENT

Are we listening yet?[1]

Marianne H. Marchand

Women in lower class and poor groups, particularly those who are non-white, would not have defined women's liberation as women gaining social equality with men since they are continually reminded in their everyday lives that all women do not share a common social status.

bell hooks (1984: 18)

While gender subordination has universal elements, feminism cannot be based on a rigid concept of universality that negates the wide variation in women's experience. There is and must be a diversity of feminisms, responsive to the different needs and concerns of different women, and *defined by them* for themselves.

Gita Sen and Caren Grown (1987: 18–19)

INTRODUCTION

Writing from the margins of feminist theory and the periphery of the global political economy, bell hooks, Gita Sen and Caren Grown send a clear message: Western[2] (white, middle-class) feminism does not represent a universal liberating force for women around the world. Western feminists' hegemonic assumption that their ideals are universal has been met with increasing resistance. To those of us who have an active interest in the field of Gender and Development these challenges should have a familiar ring. Since at least the mid-1970s, Third World women have questioned Western feminist ideals.

However, the recent arrival of the postmodernist-feminist debate at the (sub-) field of Gender and Development has taken the challenges by Third World women in new directions.[3] Various scholars have revealed (neo-)colonial discursive practices[4] within the Gender and Development literature (see Introduction/Conclusion). Among others, Chandra

Mohanty (1991b) has also identified many similarities between white, middle-class) feminist discursive practices towards women of color living in North America and Europe and Western feminist discourses toward women from non-Western societies (see chapters 4–6).[5] This has led her to question Western definitions of "Third World" women as women from non-Western societies. Instead Mohanty introduces the notion that "Third World" women constitute a political category which potentially transcends geographic borders. She defines "Third World" women as: "imagined communities of women with divergent histories and social locations, woven together by the political threads of opposition to forms of domination [sexist, racist and imperialist structures] that are not only pervasive but also systemic" (Mohanty 1991b: 4). In other words, Third World women do not only live in the South, but also at the margins of Western society.

In her article "Under Western Eyes: Feminist Scholarship and Colonial Discourses," Mohanty argues that (neo-)colonial discourses pervade the entire body of Gender and Development literature. This argument has struck a chord among Gender and Development specialists; unfortunately, her valid and valuable critique overlooks class in all its dimensions, and assumes that colonial discourse within the Gender and Development literature is simply the product of Western feminists. However, as Marnia Lazreg (1988) points out, Third World (elite) women have often contributed to the perpetuation of this discourse. In arguing that colonial discourses have been created and perpetuated by Western scholars, Mohanty implicitly denies subject agency to "Third World" women herself.[6] While not underestimating or denying the power of representation, local contributions to or attempts to change and influence dominant (feminist) colonial discourses deserve more attention. This chapter will focus specifically on these issues.

In Mohanty's broad overview of the Gender and Development literature, she makes only passing reference to scholarship by Latin American women. To fill in this gap I will explore whether and how feminist (neo-)colonial discourses have been manifested in recent literature on Latin American women, particularly the literature on Latin American women's movements. These movements clearly contradict the image of passive, voiceless and tradition-bound women. Explaining the activism and subject agency of women involved in such movements represents a theoretical challenge for feminist neo-colonial discourses. It enables me to evaluate the existence and impact of neo-colonial discourses in the field of Gender and Development in Latin America and women's efforts to influence or resist these discourses.

I will focus on two questions: (1) To what extent are neo-colonial discourses still present in the recent literature on Latin American women's movements? (2) In what ways have feminist neo-colonial discourses undergone adaptations, if any, to reflect situations where Latin American women have organized themselves and learned to speak out? The chapter is divided in three major sections. The first section discusses the concept of (neo-)colonial discourse, as developed by Mohanty and others. Subsequently, the focus shifts to the literature on Latin American women's movements, where I analyze the implicit and explicit conceptual frameworks feminist scholarship has used to explain Latin American women's movements. The important issue is whether recent feminist literature still reveals (traces of) neo-colonial discourses. Finally, I address the question of how poor working-class (Latin American) women can gain a voice and subject status in the literature on Gender and Development and actively participate in the production of knowledge about Third World women.

I focus on poor working-class women because gaining subject status is circumscribed by class as well as race and ethnicity.[7] I explore the possibility that neo-colonial discursive practices are more outspoken and also more disempowering toward underprivileged Third World women; and, in turn, these women may have more difficulty resisting these discourses. I am particularly interested in how Third World women with little formal education can actively participate in the production of knowledge about Gender and Development. What possibilities do they have, what avenues can they explore to become recognized as theorists of Gender and Development?

ON THE REPRESENTATION OF WOMEN

Both Aihwa Ong (1988) and Chandra Mohanty (1991a) have identified similar lapses into foundationalism and essentialism in (Western) feminist texts on "Third World" women (see Introduction/Conclusion). In these texts the a-historical, de-contextualized analytical category of "woman" is placed in opposition to the categories of "man" and "Third World woman," both equally a-historical and de-contextualized. In the first opposition of man versus woman, "man" is identified as oppressor and "woman" as the oppressed. In the second opposition, the category "woman" is subdivided into First World woman and Third World woman; whereby "First World woman" has been made synonymous with modern, liberated, feminist woman who is culturally superior and "Third World woman" has been relegated to an inferior position (see chapters 1 and 9).

Once this opposition has been established, a more subtle discursive move becomes apparent in which "First World woman" is identified with "woman." The opposition now reads: "(real) woman" versus (not yet modern and liberated) "Third World woman." The discursive space thus created allows differences between Western and non-Western women in terms of their experiences, their views on their roles as women, and life in general, to be *explained away* by emphasizing their difference in modernization. The implicit assumption, of course, is that when non-Western women have reached our level of modernization, they will subscribe to Western feminist ideals.[8] As a result, the Western feminist agenda can be presented (and defended) as embodying universal feminist values, and objections by non-Western women can be effectively marginalized. The fact that women from various corners of the world have challenged Western feminist ideals appears to be of no import. This is exemplified when Domitila Barrios de Chungara, in her testimony *Let Me Speak!*, relates her experiences during the International Women's Year Tribunal in Mexico City in 1975:

> I wanted to tell people all that [the sociopolitical situation in Bolivia] and hear what they would say to me about other exploited countries and the other groups that have already liberated themselves. And to run into those other kinds of problems [finding out that few people shared her interests] . . . I felt a bit lost. In other rooms, some women stood up and said: men are the enemy . . . men create wars, men create nuclear weapons, men beat women . . . and so what's the first battle to be carried out to get equal rights for women? First, you have to declare war against men. If a man has ten mistresses, well, the woman should have ten lovers also. If a man spends all his money at the bar, partying, the women have to do the same thing. And when we've reached that level, then men and women can link arms and start struggling for the liberation of their country, to improve the living conditions in their country. That was the mentality and concern of several groups, and for me it was a really a rude shock. We spoke very different languages, no?
>
> (1978: 199)

Barrios de Chungara continues her account of her experiences at the Tribunal by describing how she felt alienated because her and "her people's" interests were not reflected in the agenda, which, according to her, primarily reflected the concerns of Western feminists and upper-class Latin American women, two groups with which she does not identify herself. Barrios de Chungara's challenging of the privileging of Western feminist concerns is described in her encounter with a leading US feminist, Betty Friedan:[9]

> [Betty Friedan] and her group had proposed some points to amend the 'World Plan of Action.' but these were mainly feminist points and we didn't agree with them because they didn't touch on some problems that are basic for Latin American women. Betty Friedan invited us to join them. She asked us to stop our 'warlike activity' and

said that we were being 'manipulated by men,' that 'we only thought about politics,' and that we'd completely ignored women's problems, 'like the Bolivian delegation does, for example,' she said.

(1978: 201–2)

This account illustrates the disciplining of Third World feminists and feminism; a set of discursive practices were employed to marginalize her interests. These agenda-setting efforts were accompanied by more overt forms of marginalization, such as limiting her access to the microphone, thus literally silencing her voice. However, the silencing was only temporary, because Barrios de Chungara recovered her voice by giving her testimony and recounting her encounter with Betty Friedan.

Clearly, the context of women's lived realities, i.e. place, informs women's perceptions about justice, equality and women's issues. However, in Mexico City Western feminist ideals could be presented as universal, precisely because the discursive space had been created by representing Third World women as the inferior "other." This representation disallows differences among women from the South and denies them (subject) agency. In the remainder of this chapter I will discuss whether a (similar) essentialist Western feminist voice is still present in the literature on Latin American women's movements.

LATIN AMERICAN WOMEN'S MOVEMENTS AND FEMINIST NEO-COLONIAL DISCOURSES[10]

To development theories generated in the North the geographical area of Latin America has presented somewhat of a (theoretical) challenge. As far as modernization theory is concerned, Valenzuela and Valenzuela (1981: 17) point out that "specialists on Latin America failed to contribute important theoretical efforts to the field," because Latin America did not quite fit the picture painted by modernization theorists. For instance, unlike other parts of the "Third World" most of the Latin American countries gained independence during the 1820s. In addition, "Latin America's close (particularly cultural) ties to the West, made it more difficult, by contrast with Asia and Africa, to point to obvious differences with the European experience" (Valenzuela and Valenzuela 1981: 17). Latin America thus posed a serious challenge to modernization theory. This challenge was overcome by a discursive move whereby traditional values became synonymous with a "Mediterranean ethos," which explained "backwardness" in Latin American societies in terms of the modernization theory.[11]

This discursive process has implications for the representation of Latin American women in the Gender and Development literature. The region's close connection with the West has affected feminist discursive practices for the region. "Latin American" women occupy a special place within feminist neo-colonial discourses about Third World women. More specifically, the problem faced by feminist scholarship has been: How to treat Westernized, middle-class, university-educated Latin American women as well as women who actively participate in women's movements, since both groups defy the classification of passive, voiceless and tradition-bound "Third World" women?

Recently, feminist scholarship has turned its attention to the relations between the state and Latin American women's movements, specifically their role in transition politics. In reviewing some of this literature (Alvarez 1990; Bourque 1989; Jaquette 1989; Molyneux 1985; Navarro 1989; Safa 1990) it should be noted that these authors are very careful to emphasize the heterogeneity of Latin American women's movements. Yet, within this literature some interesting patterns emerge. First, virtually all authors make a distinction between feminist and feminine movements. For example, according to Sonia Alvarez:

> Whereas feminist organizations focus on issues *specific* to the female condition (i.e. reproductive rights), feminine groups mobilize women around gender-related issues and concerns. The cost of living, for example, is one such issue. The sexual division of labour in most societies holds women responsible for managing family budgets and allocating family incomes to provide for basic necessities. Women, then, may organize to protest the rising cost of living because inflation undermines their ability to adequately feed, clothe, or house their families.

(1990: 25)

The distinction between feminist and feminine movements can at least be traced back to Maxine Molyneux's article "Mobilization without Emancipation? Women's Interests, the State, and Revolution in Nicaragua" in which she juxtaposes practical and strategic gender interests – the former being pursued by feminine movements, the latter by feminist groups (1985: 232–3). However, the feminist–feminine dichotomy is not just a social construct imposed by Western feminists; it is used by Latin American women themselves. For instance, most middle-class, university-educated women who are actively involved with "women's issues" consider themselves feminist, whereas poor, working-class women often reject the "feminist" label (Alvarez 1990; Jaquette 1989; Schmink 1981).

A close reading of the literature reveals yet another dimension of the feminist–feminine dichotomy. Some take this dichotomy as a given and use it as *the starting-point* for their analyses. At the same time their own research reveals that the situation may not be as clearcut (read: dichotomous) as the feminist–feminine distinction makes us believe. Helen Safa (1990), for instance, points out that practical gender interests can and

61

often do lead to strategic interests. Sonia Alvarez (1990), in turn, discusses how Brazilian "feminist" groups actively sought to create cross-class alliances with "feminine" groups by formulating a common agenda that included gender-specific as well as gender-related issues.

Lastly, the distinction between practical and strategic gender interests ultimately also informs authors' assessments of the failures and successes of Latin American women's movements and their futures. Jane Jaquette argues that feminine groups are more likely to be subject to (traditional) clientelistic practices than feminist movements (1989: 191–2). According to Helen Safa, only a movement away from gender-related (feminine) concerns towards an increased awareness of gender-specific (feminist) issues will keep Latin American women politically mobilized (1990: 366). However, this reliance on the feminist–feminine dichotomy circumscribes and constrains their analyses.

Fortunately, the dichotomization of practical and strategic gender needs and of feminine and feminist movements has recently come under closer scrutiny.[12] Sarah Radcliffe and Sallie Westwood reject it as a starting-point. They write that:

> the distinction is also problematic. It suggests that there is a simple dichotomy between 'practical' and 'strategic' gender interests which can be aligned with notions of the public and the private as spheres of interest for women; this, as we suggest, may be helpful for organizing commonsense but does not provide a theoretical base for understanding women as political subjects and actors. We would also want to suggest that it, too, has a universalizing quality which is located with a linear view of progress founded upon the post-Enlightenment account of movement towards a goal as part of a grand narrative of rational progress towards a better world. . . . Such meta-narrative suggests a hierarchical relationship between practical and strategic gender interests such that women, in order to progress, must move from one to the other.
>
> (1993: 20)

Despite this pertinent intervention, the dichotomization of practical and strategic gender interests continues to be widely present in the Gender and Development literature (see Moser 1993).

The reliance on the feminist–feminine dichotomy raises important questions. First, is this dichotomy an expression of feminist neo-colonial discursive practices? Second, does the dichotomy misrepresent the lived material realities of Latin American women? At this point, it is instructive to return to bell hooks. Writing from the vantage point of an African-American woman in the United States, hooks argues that feminism should fight against sexism, as well as racism and classism. In her words,

One of the basic differences in perspective between the bourgeois woman and the working class or poor woman is that the latter knows that being discriminated against or exploited because one is female may be painful and dehumanizing, but it may not necessarily be as painful, dehumanizing, or threatening as being without food or shelter, as starvation, as being deathly ill but unable to obtain medical care. Had poor women set the agenda for the feminist movement, they might have decided that class struggle would be a central feminist issue; that poor and privileged women would work to understand class structure and the way it pits women against one another.

(1984: 60–1)

In this quote, hooks convincingly argues that the gender-related survival needs of poor women of color should be as much part of feminist theory as the struggle for reproductive rights (see chapters 2, 4–6).

Hooks' perspective has important implications for the discursive practices of feminist scholarship on Latin American women's movements. If one accepts her definition of feminist theory, the feminist–feminine dichotomy reveals a dual hierarchized opposition, which structures the discourse about Latin American women's movements similarly to the discursive role played by the Mediterranean ethos in modernization theory; in other words, the feminist–feminine dichotomy amounts to an adaptation of feminist neo-colonial discourses which enables Western feminist scholars to reify the superiority of Western feminism and consequently to neutralize the "problem" of Latin American women's movements. They thus send an important message: only a very small percentage of Latin American women are "real" feminists and can thus be considered modern women (like us!). As it so happens, many of these women were in exile in Western Europe or the United States during the 1970s, where they had their first encounter with Western feminism (Alvarez 1990). At the same time, the feminist credentials of the large majority of Latin American women involved in social and political movements is questioned since their actions are seen merely as extensions of the *traditional* female or domestic sphere.

A strong case can be made that the feminist–feminine dichotomy distorts our representation of Latin American women's movements. At the same time, it should be recognized that working-class women in Latin America quite regularly use this dichotomy as a source of empowerment. Within societies where machismo and its counterpart marianismo have (until recently) generated clearly defined expectations about gender roles, women may need to justify their action(s) as an extension of their duties and roles as women. In this way they create their space within the public sphere, while denying the feminist and political nature of their activities (Navarro 1989; Pires de Rio Caldeira 1990). A straightforward statement about women's "feminist" actions could easily arouse male opposition as well as women's anxiety about bridging socially ascribed gender roles. The eradication of the feminist–feminine dichotomy might, under certain circumstances, lead to a different form of neo-colonial discourse, denying local women the possibility to create their own political and discursive space. The difference between the discursive imposition of a feminist–feminine dichotomy and local women's interpretations of this dichotomy resides with the issues of empowerment and representation. In the latter case, local women have a voice in their own representation, and can use the dichotomy for their own empowerment.

Of course, there is no reason why feminist theory on Latin American women's movements necessarily has to follow the direction(s) taken by US feminist theory. While there seem to be some grounds for keeping the distinction between feminine and feminist issues, it is imperative to avoid any undue hierarchization between the two. Therefore, I would argue that we should place practical and strategic gender interests on a continuum rather than pose them as a dichotomy. This construction seems more in tune with actual practices, whereby gender-specific and gender-related needs are intertwined (Radcliffe and Westwood 1993).

In sum, we need to "strike a balance" in theorizing about "Third World" women. The differences and diversity among Latin American women's movements must be acknowledged without adopting a rank ordering of women's issues whereby the struggle for reproductive rights is classified as "real feminism" while neighborhood improvements and survival needs are considered "fake feminism." Such a move would merely reinforce an adaptation of neo-colonial discursive practices by reiterating the superiority of Western feminist ideals.

In the next section, I will explore how poor, working-class women in Latin America might influence the production of feminist theory in the area of Gender and Development. Because of their position in the sexual division of labor, many Latin American women involved in grassroots (i.e. feminine) movements find themselves at the margins of the production of feminist theory. This leads to alienation from feminist issues and scholarship. Yet, feminist scholars acknowledge the importance of many strategic (i.e. feminist) gender interests. Clearly, we need to find ways in which poor, working-class women's feminine concerns can actively participate in the production of feminist theory.

TESTIMONIES AS POSTCOLONIAL PRODUCTION OF KNOWLEDGE?

In order to challenge dominant Western feminist discourse on non-Western women, we need to create discursive spaces which will allow the voices of Latin American and other Third World women to be heard. As Ong argues, we need to "recognize other forms of gender- and culture-based subjectivities, and accept that others often choose to conduct their lives separate from our particular vision of the future" (1988: 90). This can be done by contextualizing and historicizing Third World women's practices and experiences. In this section, I will explore the possibility that testimonies can be used to create a discursive space which would allow Latin American women from poor, working-class backgrounds to speak out and participate in the production of feminist theory.

Various academic disciplines use (oral) life histories as a research method, a source of information (especially anthropology and history) or a specific literary genre (literary studies). In this chapter I distinguish between life histories which "refer to the generic category of recorded participant narrative" and testimonies which "denote[s] life histor[ies] of a political, denunciatory nature" (Marchand 1994). The role of life histories and testimonies has been much debated within anthropology and Latin American studies.[13] Interestingly, within these debates certain issues of importance for the present analysis have been marginalized or silenced.

Life histories as ethnographic texts have been part of anthropology at least since the 1920s (Langness and Frank 1981; Watson and Watson-Franke 1985). Now-famous anthropologists such as Margaret Mead and Oscar Lewis have contributed to this long tradition. Despite all this, life histories have always occupied an ambiguous position within the discipline. As Vincent Crapanzano points out:

> the life history has been somewhat of a conceptual – and emotional – embarrassment to academic anthropology and has remained on the periphery of the discipline . . . [it] is more "literary" than "scientific" – and yet more "scientific" than "literary." It mediates, not too successfully, the tension between the intimate field experience and the essentially impersonal process of anthropological analysis and ethnographic presentation.
>
> (1984: 954)

Recently, however, anthropologists have rediscovered the life history as respectable ethnographic text. This has occurred within the context of a self-reflective and self-critical moment, influenced by contemporary literary criticism. This self-criticism is directed at what can be considered the very core of anthropology, i.e. the (possibility of) representing other cultures (Clifford and Marcus 1986; Marcus and Fischer 1986). Taking the critique of colonial discourse[14] to its logical extreme undermines anthropology itself, because it eliminates the discipline's legitimizing discourse. In other words, what would anthropology be without the representation of other cultures? Fortunately (for anthropologists), scholars who formulated the critique of colonialism in the first place, also rescued the discipline from its demise in a set of inventive discursive moves. For the anthropologist facing the "problem" of representation, at least two options are available. The first is laid out by Marcus and Fischer:

> as [modern anthropologists] have written detailed descriptions and analyses of other cultures, ethnographers have simultaneously had a marginal or hidden agenda of critique of their own culture, namely, the bourgeois, middle class life of mass liberal societies, which industrial capitalism has produced.
>
> (1986: 111)

Besides the option of "representing other societies" in order to critique one's own culture, experimental anthropologists have begun creating ethnographic texts in the form of a dialogue (Dwyer 1982) or a polyphony of voices.[15] As Clifford (1986: 15) notes, the attraction of the second option is that: "once dialogism and polyphony are recognized as modes of textual production, monophonic authority is questioned, revealed to be characteristic of a science that has claimed to *represent* cultures."

At first glance, the life history provides the perfect model for this new ethnographic text, "because it allows the "anthropological subject" to speak herself and to represent her own society" (Marchand 1994: 139). Yet, after having introduced the life history as a model for the experimental anthropologist, Clifford subsequently casts doubt on the viability of the project:

> Once 'informants' begin to be considered as co-authors, and the ethnographer as scribe and archivist as well as interpreting observer, we can ask new, critical questions of all ethnographies. However, monological, dialogical, or polyphonic their form, they are hierarchical arrangements of discourses.
>
> (1986: 17)

The ethnographer will still hold an authoritative position, because of the dynamics involved in the interview process itself and the fact that she has the power of production. For instance, the ethnographer 'selects the subject' and directs the interview by asking questions. These questions are based on what the ethnographer deems important. The ethnographer's ultimate authority lies in her power of production. "When creating her text, the ethnographer engages in contextualizing/representing her informants' voices and through these actions either validates or delegitimizes them" (Marchand 1994: 139). Facing the dilemmas of this post-anthropological moment, the best one can hope for are partial truths:

> Even the best ethnographic texts – serious, true fictions – are systems or economies, of truth. Power and history work through them, in ways their authors cannot fully control. Ethnographic truths are thus inherently *partial*–committed and incomplete.
>
> (Clifford 1986: 7)

Under these circumstances, it is important to contextualize the life history, i.e. explain how a particular project came into existence (Behar 1990; Dwyer 1982; Patai 1988; Watson 1976).

In contrast, students of Latin American literature have been concerned with the authorship of testimonies:

> It could be argued that the anthropologists' questions about representation have been reformulated by literary scholars into questions of authorship. The literary question, Who is the real author? overlaps with the anthropological question, Who is speaking/Whose voices are we hearing?
>
> (Marchand 1994: 139)

Within Latin American literary studies this problem has been partially resolved by either assuming single or multiple authorship (Behar 1990; Beverley and Zimmerman 1990; Jara and Vidal 1986).

As with the role of life histories or ethnographies in anthropology, discussions about testimonies touch the foundations of literary studies as a discipline (Beverley and Zimmerman 1990; Jara and Vidal 1986). Supposedly, the field of literary studies has as its object the study of (literary) fiction. Yet testimonies have emerged to denounce oppressive lived material realities. The discursive dilemma that emerges from a literary perspective is clear: "if one considers testimonies to be truthful, then they can only be *nonliterary* documents; on the other hand, if one assumes that testimonies are literature, they *have to be* fictional" (Marchand 1994: 139). This (hierarchized) opposition finds its roots in the dichotomy of science versus literature (or the humanities) which culminated in the Enlightenment project (Marchand 1994). As Clifford notes the dichotomy provides the foundation of Western science:

> Since the seventeenth century, they suggest [Michel Foucault, Michel de Certeau, and Terry Eagleton], Western science has excluded certain expressive modes from its legitimate repertoire: rhetoric (in the name of 'plain,' transparent signification), fiction (in the name of fact), and subjectivity (in the name of objectivity). The qualities eliminated from science were localized in the category of 'literature.' Literary texts were deemed to be metaphoric, and allegorical,

composed of inventions rather than observed facts; they allowed a wide latitude to the emotions, speculations, and subjective 'genius' of their authors.

(1986: 5)

Beverley and Zimmerman (1990) posit that testimonies are fact and fiction, and thus challenge the discursive boundaries of the field of literary studies.

The most interesting part of these debates is not their similar concern(s), but rather the question(s) that have been ignored. Scholars within these two fields of study have focused on the appropriate standards for data gathering, representation and authorship; issues that reflect the disciplinary concerns of anthropology or literary scholarship. Little attention is paid to why people volunteer to become informants. I contend that this negligence is a very subtle form of neo-colonial discourse by precisely *those who have formulated critiques of neo-colonial discourse,* because it still implicitly denies Latin American women subject status as "real, but different women." This last discursive move is grounded in the notion that Latin American women, who decide to give their testimony, "do not actually have the independence or autonomy to make a decision about whether or not to participate" nor do they have the "ability to steer the interview with the anthropologist" (Marchand 1994: 140). In short, while I recognize certain power dynamics between anthropologist and informant when a life history/testimony is being created, I believe that critiques of neo-colonial discourse themselves ultimately rest upon a neo-colonial discursive move if they do not take into account the decision to offer a testimonial.

It is imperative to overcome the assumption of the "innocent or defenceless native" in order to consider the potential(ities) of testimonies in the production of knowledge about Gender and Development. In other words, testimonies might present a venue for women, who are politically/culturally/educationally/socially/economically marginalized, to have their development practices and knowledge impact on the field of Gender and Development. Having said this, we should always keep in mind that ultimately testimonies construct the subject as well (Goetz 1991; see Introduction/Conclusion).

Using various testimonies by Latin American women as examples (Benjamin 1989; Burgos-Debray 1984; Massolo 1992), we can glean some insights into their potential contributions to the field of Gender and Development. Since testimonies are about telling a life story, most start with an elaborate account of a woman's life from birth till present. This information is not merely background information; it can serve to contextualize Gender and Development issues. For instance, Elvia Alvarado talks about the profound influence of machismo on her life:

her father did not allow her sisters to go to school, her mother left Elvia and her siblings to live with a man and her brother threatened to kill her when she got pregnant (Benjamin 1989: 2–3). Similarly, the ways in which women become empowered and join popular organizations is contextualized in testimonies. Rigoberta Menchú, for instance, explains her move towards politicization:

> My dreams came true when we started organising. Children had to behave like grown-ups. We women had to play our part as women in the community, together with our parents, our brothers, our neigh-bours. We all had to unite, all of us together. We held meetings. We began asking for a community school. We didn't have a school. We collected signatures. I was involved in this. . . . At the end of 1977, I decided to join a more formal group–a group of peasants in Huehuetenango. . . . And yet, I still hadn't reached the rewarding stage of participating fully, as an Indian first, and then as a woman, a peasant, a Christian, in the struggle of all my people.
>
> (Burgos-Debray 1984: 120)

These testimonies not only contextualized Gender and Development issues, they also reflect their authors' views on such issues. Without exception, Elvia Alvarado, Rigoberta Menchú and Señora Aurora[16] take a holistic approach to Gender and Development questions. This approach is most striking in Elvia Alvarado's testimony. She makes connections among such issues as: (a) economic growth; (b) redistribution of land; (c) women's concerns about domestic violence and alcoholism, reproductive rights, health, etc.; (d) militarization; (e) democratization; and (f) the role of the United States in Honduras. For Elvia Alvarado development is not only a multifaceted concept which needs to be addressed in its entirety, but also a highly "gendered" concept. This follows from her definition of development (objectives): "we have to build a society where everyone has the right to live a decent life" (Benjamin 1989: 27). "But if we really want to build a new society, we have to change the bad habits of the past. We can't build a new society if we are drunks, womanizers, or corrupt" (1989: 55). In other words, the "right to a decent life" is inextricably linked to the elimination of machismo.

For Rigoberta Menchú, development is very much tied to a politics of identity. Interestingly, she doesn't use the term development in her testimony. While she considers improving the living conditions of Guatemala's poor to be important, she argues that it cannot be separated from granting cultural autonomy to the Indians:

> That is my cause. As I've already said, it wasn't born out of some-thing good, it was born out of wretchedness and bitterness. It has

been radicalized by the poverty in which my people live. It has been radicalized by the malnutrition which I, as an Indian, have seen and experienced. And by the exploitation and discrimination which I've felt in the flesh. And by the oppression which prevents us performing our ceremonies, and shows no respect for our way of life, the way we are.

(Burgos-Debray 1984: 246–7)

Throughout Señora Aurora's account one can identify a variation on the feminist concept that "the personal is political." For her, development is personal. Families in her squatter settlement Ajusco have had to struggle (with the environment, landlords, local government and project developers) for every brick, cobblestone and drop of piped water. Indeed, neighborhood organizing could not have been done without the women. In order to continue to improve their living conditions (i.e. develop), women will have to improve their own situations and address the problems related to machismo (Massolo 1992).[17]

In sum, a quick perusal of these three testimonies reveals a goldmine of information, ideas and knowledge about Gender and Development issues. For all three women, development means more than the improvement in material living conditions. Contextualizing Gender and Development issues, formulating a holistic approach to problems, and accepting the personal character of development are major themes in these testimonies. They also reveal the struggles of poor women to overcome the discursive explanations of their lives as poor, helpless and ignorant, and the capacity of women, whatever their education/position to define their lives, albeit frequently in circumstances not of their own choosing.

CONCLUSION

In this chapter, I address the connections between (neo-colonial) representations of "Third World women" and the production of knowledge about Gender and Development. Neo-colonial discourse is multilayered and firmly embedded in Western Enlightenment thought, in particular Western concepts about knowledge and science. The question arises how women from the South can undermine, resist and transform this power/knowledge nexus and become full participants in the Gender and Development enterprise. In an attempt to answer this question, I have explored the possibility that Latin American women can gain a voice and subject status through their testimonies.

In the literature on Latin American women's movements hierarchies between Western(-educated) liberated feminists and marginalized Latin American women are maintained through the dichotomization of practical and strategic gender interests. The imposition of this (discursive) hierarchy cannot be justified on the grounds that Latin American women sometimes use the distinction for their own ends. The discursive imposition of the practical/strategic gender dualism is an act of disciplining and hierarchization; utilizing the distinction to legitimize one's political mobilization is emancipatory in nature.

Reading testimonies against this background, it is clear that they provide one venue for (marginalized) Latin American women to participate in the production of knowledge about Gender and Development. Although the issue of representation in testimonies remains problematic, both because of the partial construction of the subject and the intervention of the ethnographer, I think that it is important to recognize that women have agreed to give their testimony and have approached them in a very utilitarian fashion. These testimonies are a means by which women can conduct their struggles to resist and delegitimize dominant discourses. Indeed, testimonies are probably the only means for (marginalized) Latin American women to conduct their struggle(s) at the level of the production of knowledge. Through testimonies they can impart their views, practices and expertise on Gender and Development. They can situate their life experiences as a form of resistance to the dominant discourse, and remind the reader/practitioner that for most women, development is a highly gendered process which starts at home. Thus, not only is "the personal political," "development is personal" as well.

NOTES

1 I wish to thank the series editors, as well as my friend and co-editor Jane Parpart, for their insightful comments on an earlier version.
2 The terms West and East are simultaneously embedded in Cold War thinking and earlier colonial notions about differences between the Orient and the Occident. Because of recent geopolitical changes, in particular the ending of the Cold War, the term West is increasingly replaced by North(ern).
3 For an excellent overview see Nicholson 1990.
4 I use the term neo-colonial discourse(s) to indicate that politically we are living in the post-colonial era although economically, socially and culturally there are still colonial structures present.
5 For an elaboration of this idea see the Introduction/Conclusion and Marchand 1994.
6 See below.
7 This focus is not meant to contribute to the so-called victimization of women; rather I assume that most, if not all, women have knowledge about different forms of oppression, i.e. gender, class, ethnicity, etc., which they might want to share with other women. However, poorer women in the Third World have fewer possibilities for such sharing.
8 See the statement by Safa (below) as illustration.
9 Of course, Betty Friedan was not the sole representative of Western feminism; she is, however, a leading US feminist.

10 For a related discussion, see Marchand 1994.
11 The term "Mediterranean ethos" refers to the notion that people of Iberian heritage tend to be lazy, anti-democratic (i.e. infatuated with hierarchical types of organization), anti-entrepreneurial and militaristic.
12 See Radcliffe and Westwood's introduction (1993) and Nikki Craske's contribution on "Women's Political Participation in *Colonias Populares* in Guadalajara, Mexico."
13 See the two special issues "Voices for the Voiceless" of *Latin American Perspectives*, Summer and Fall, 1991. I am not entering into the debates among historians about oral histories, because they have, in contrast to anthropologists, focused less on Gender and Development and non-Western cultures.
14 Mohanty (1988; 1991a) and Ong (1988) apply this concept to feminist writings on Women and Development.
15 The idea of a polyphony of voices is based on Bakhtin's concept of 'heteroglossia' (1981).
16 Señora Aurora has testified about her life in a squatter settlement in Mexico City to Alejandra Massolo. Her testimony and those of a few other women form the bulk of Massolo's book on women in urban movements.
17 In her testimony Señora Aurora discusses the policy of another urban popular organization in Monterrey where the women go together to a husband if he doesn't allow his wife to participate. Also, if a husband beats his wife, he will be publicly criticized and expelled from the organization (Massolo 1992: 198).

Part III

WOMEN OF COLOR: REPRESENTATION AND THE SOCIAL CONSTRUCTION OF IDENTITY

Historically, the literature on Gender and Development rarely (if ever) takes into account women of color living in industrialized nations. The authors contributing to this section of the book set out to (partially) fill this gap. The contributions by Julia Emberley, Catherine Raissiguier and Suresht Bald serve to explode some of the premises upon which the Gender and Development literature (and general development literature) is based.

The emphasis on continuities between North and South is not entirely new. Various policymakers, academics and NGOs have acknowledged the presence of "Third World enclaves" within the First World. In the past, however, this has not led to theoretical reconceptualizations of North and South. On the contrary, development theory has continued to produce and reproduce (hierarchized) dichotomies such as North/South, First World/Third World, developed/underdeveloped and modern/traditional. Indeed, one could argue that development theory on the one hand creates the South as inferior "other," while at the same time owing its existence to the construction of these hierarchical dichotomies. As Deborah Johnston points out:

> So far, theories of development (both conventional and more "radical") have attempted to uncover the truth about underdevelopment – what it is, why it is, and what may be done to remedy it. But in their analyses, they presuppose the very distinction they seek to examine. They accept as natural and necessary the distinction between First and Third Worlds, between core and periphery, between developed and underdeveloped.
>
> (1991: 161)

Thus by exploring the continuities and discontinuities between (neo-)colonial discourse(s) about "Third World" women and similar discourses on First Nation and Diaspora women, the three chapters in this section challenge the modernist foundations upon which (gender and) development practices and theories have been built, including the naturalized (hierarchical) dichotomy between the "modern North" and the "traditional South." Another reason why the analysis of women of color deserves attention in a book on feminism and development issues is that the writing of such authors as bell hooks, Marnia Lazreg and Chandra Mohanty have inspired feminists in the South. This under-scores at yet another level the continuities between North and South, between "developed" and "developing."[1]

Although it is important to challenge the Gender and Development literature for its silence on women of color, a note of caution is in order. The attempt to uncover and problematize development theory's natur-alized and "necessary" dichotomy between traditional and modern or core and periphery at the international level should not result in the (re)creation of a similar dichotomy between modern, white middle-class feminists and traditional, poor women of color within the industrialized world. In their contributions, Bald, Emberley and Raissiguier try to avoid this. They focus on how dominant discourses produce and repro-duce women of color, and thus circumscribe their lived material realities. They also show how these women are actively involved in constructing their own identities within the dominant discourse(s) as well as at the intersection of the dominant discourse(s) by the former colonizers and the discourses of their own groups. In other words, simplistic dichoto-mies between "modern," white middle-class Western women and tradi-tional, poor women of color negate the complex identities of both groups.

The authors highlight various themes that are central to the repre-sentation of women of color and their social construction of identity. These themes include: (1) the question of sexuality, especially for young women of color; (2) the dualism or dichotomy that exists between the discourses of the group to which the women "belong" and that of the dominant (colonial) group; and (3) the issue of locale or context. Obviously, these issues are not just separate items; all three chapters clearly show how they interact or reinforce each other. The chapters also highlight possible continuities and discontinuities between the experi-ences of women of color (in the North) and women in the South.

In her chapter, Emberley raises the question of sexuality, in particular sexual violence, most poignantly. She analyzes Ann Wheeler's film *Loyalties*, which exhorts the liberal feminist ideal of solidarity among all women by using the rape of a young Native woman as a rallying point for white middle-class and Native women. Emberley's critical analysis reveals the problematic relationships between these groups. According to her, contradictions between the ideal of liberal feminism and the daily experiences of Native women leave very little room for solidarity. These contradictions are brought about by the dominant colonial discourse which is inscribed onto the body of the "Native woman." In other words, sexual violence (toward Native women) and colonial discourse reinforce each other: the territory of indigenous people in Canada becomes equated with a (Native female) body and this "feminized" space is then available to be conquered. First Nations are being transformed into Native/Woman/Body. Once the duality between the (male) conqueror and (female) conquered has been established, it circumscribes and encapsulates the relationship between the white middle-class woman and the Native woman. Emberley argues "that the post-colonial condition of Lily's [the European woman's] freedom is gained at the expense of Leona's [the young Native woman's] victimized body." In the end, contradictions between anti-imperialist and feminist projects make the latter's quest for global solidarity near to impossible. According to Emberley, "the site of struggle to articulate the significance of these movements [new social movements] becomes, however, Leona's body; a Native woman's body that is sexualized and naturalized as a legitimate site for the resolution of social, political, and economic conflict." Thus Emberley's analysis on women's bodies and sexuality reminds us how these issues have broader implications for questions of Gender and Development.

In Emberley's chapter the question of sexual violence prevents Native women from speaking for themselves and constructing their own identity. Similarly, Raissiguier sees sexuality as an important element in the identity formation of Algerian high school girls. For these girls school is a "social space, . . . at once a contested terrain within which these young women's identities are being constructed and a powerful tool for their own identity formation." It is also within the context of school that the young Algerian women contrast their own sexuality, which is tightly controlled by the male members of their families, with the sexuality of promiscuous French classmates. Yet, these Algerian women have not entirely internalized the duality of the "liberated French girl" versus the "traditional Algerian girl," which is being reproduced in the discourses of the Algerian community and of the (former) French colonizer. In an attempt to construct their own identity, the Algerian high school girls neither fully resist nor accommodate to this duality – thus showing that they have freedom and are not without agency. In this process of identity construction their school and education become very important, especially when it concerns escaping the gaze of their male family members. As with their (young) Asian counterparts in Britain, these Algerian women try to make the most of being "suspended between

two cultures/cultural discourses." In many respects they consider themselves better off than young women living in Algeria. The chapter also
emphasizes the complex nature of education. It reminds us that education is not just a simple gender-neutral development tool, which serves to
(unidirectionally) transmit knowledge and training to people in the
South. Rather schools and education are, among other things, intricately linked to the production and reproduction of young women's
identities.

In the chapter by Bald the question of sexuality also appears, although
not as prominently. Sexuality is especially important for the young Asian
women interviewed by Bald. Their families tend to place much pressure
on them to accept arranged marriages. For some this has resulted in
abusive relationships. However, it is also clear that living in Britain (and
thus some distance from South Asia or East Africa) makes it possible
sometimes to escape from an abusive relationship by going to a lawyer,
getting council housing etc. In sum, the question of sexuality appears in
all three chapters as an important issue in the lived material realities of
women of color. It is also clear from the accounts that the question of
sexuality does not appear in isolation; it is closely tied to the existing
duality between the dominant discourses of the (former) colonizers and
the marginalized discourses of these women's ethnic communities.

In the three chapters the question of duality between dominant
discourse and marginal discourse is treated differently. Emberley, for
instance, identifies a hierarchy between the anti-imperialist and (liberal) feminist discourse which accounts for only a fragmented identity
of Native women. According to her, the unbroken connections between
colonial discourse and liberal feminism circumscribe the representation
of Native women and the possibility of Native women to have a voice
within the two dominant discourses. Raissiguier, on the other hand,
places the dominant (liberal) French discourse of assimilation versus
the marginal discourse of the Algerian community. In the attempt at
constructing their identity, the Algerian high school girls she interviewed
find themselves "suspended" between these two opposing discourses.
Finally, Bald accepts almost as given dominant British, racist discourse
which denies differences among members of the "Asian community."
However, within the marginalized discourse of this constructed "Asian
community" Asian women actively engage in constructing their identity
and agency. In so doing, they are sometimes also "gnawing" at the edges
of the dominant British discourse, thus undermining the latter by
revealing its contradictions. This is in particular true for community
activists whose interests and activities prevent them from staying within
the confines of the marginalized community's discourse.

The duality between dominant and marginal discourses might be a
contradiction more acutely experienced by women of color living in the

North than for women in the South. However, it is important to consider whether comparable dualities between dominant and marginal discourses circumscribe Third World women's agency and (re)production of identity.

The theme of context, locale or space is by all three authors used in many different ways. In the three contributions, the theme has not only received a territorial connotation, but has also been used to identify social, emotional/psychological and discursive dimensions. The use of context or locale is most developed by Emberley in her discussion about editing techniques and by her focus on the equation body = territory. Raissiguier uses both the territorial and social dimensions of space: (a) the school buildings constitute a territory separate from the home, where Algerian high school girls can engage in identity construction (for instance through their conversations with each other); while (b) education provides a certain social "room for manoeuver," because these young women have escaped the pressure to marry at a very early age and because it might help them move up the social ladder.

Related to the theme of context or locale is the idea of distance which also has social, emotional or geographical dimensions. One way in which this comes out is the space or distance between authors and subject matter of their chapter. This can be expressed for instance in the extent to which they identify themselves with First Nation or Diaspora women. Of related interest is the issue, from what vantage point are the authors writing: Emberley's analysis is grounded in cultural studies, while Bald and Raissiguier approach their respective analyses from the social sciences. Obviously, these different approaches influence the analysis and presentation of issues. In juxtaposing these approaches it is not their differences that are striking, but the extent to which they are complementary and raise comparable issues.

The importance of this section is not (just) to raise new issues for the Gender and Development literature to incorporate, but to fundamentally challenge the ontological and epistemological foundations of development literature in general. To adapt Edward Said's observation about Orientalism to (Gender and) Development literature:

> [it] is a style of thought based upon an ontological and epistemological distinction made between ["the First World"] and (most of the time) ["the Third World"]. Thus a very large mass of writers, among whom are poets, novelists, philosophers, political theorists, economists, and imperial administrators, have accepted the basic distinction between [North] and [South] as the starting point for elaborate theories, epics, novels, social descriptions , and political accounts concerning the [South], its people, customs, "mind," destiny, and so on.
>
> (1979: 2–3)

The distinction between North and South influences, for instance, the way the issue of women's sexuality can be raised. In the past, the question of sexuality has been linked to "family planning": the traditional, Third World woman has been primarily represented in her reproductive role of mother and "breeder," responsible for the "overpopulation" in the South. Not surprisingly, the solution to this development problem has also been found within the parameters of the modern–traditional dichotomy: through education these women will become acquainted with modern values and their fertility rates will drop as a result. The North/South distinction makes it very difficult, if not impossible, to raise the question of sexuality in the Gender and Development literature in a way that mirrors the three contributions to this section. Reproductive rights of women living in the South are made subordinate to other (more important?) objectives. This example suggests the need for new ways of thinking about development, ones that can escape the North/South divide and incorporate the experiences/ knowledge of minority women in the North.

NOTE

1 Similar continuities among First Nation women, women living in Diaspora and "Third World" women have been explored in C. Mohanty, A. Russo and L. Torres (eds) (1991) *Third World Women and the Politics of Feminism*, Bloomington, IN: Indiana University Press.

4

THE CONSTRUCTION OF MARGINAL IDENTITIES

Working-class girls of Algerian descent in a French school

Catherine Raissiguier

INTRODUCTION AND THEORETICAL DEBATES

This chapter analyzes the processes of identity formation among working-class female students of Algerian descent in a French school. It argues that the school – as a social space – is at once a contested terrain within which these young women's identities are being constructed and a powerful tool for their own identity formation.

Indeed, the public (state) school, because it is a secular space, brings different groups into contact with one another materially and symbolically and pulls competing and conflicting discourses onto a shared terrain. This chapter explores how a particular group of young women actively engage in a work of self-definition within and against these competing discourses.

The first half analyzes how the Algerian community and French mass media create a potent discursive link between the young women and school. The second part explores the ways in which the young women construct themselves within the school in relation to others – in particular to French women in their age-group and similar class background and men within their own community.

As I pose the seemingly simple research question "How do young women construct themselves in relation to others in a given context?", the raucous, explosive, troubled debates that have emerged at the margins of feminist and postmodern scholarship come to mind. How does one frame a non-essentialist analysis of the construction of subjectivity that allows for agency while still recognizing the existence of material and discursive boundaries within which the agent is constituted?

Foucault's work on discourse, power, and knowledge informs my own attention to language in the process of identity formation. However, it is also in the spirit of Audre Lorde, Trinh Minh-Ha, bell hooks and

79

Patribha Parmar – to name just a few – that I have included in my analysis a discursive layer. Indeed, one of the greatest Master's tools is language and its ability to represent the other(s). While it is crucial to recognize that discourses always shape the ways in which we can apprehend reality, it is as important to locate these discourses in the lived, historical and material situations in which they circulate. Dorothy Smith calls for a similar grounding of our discursive analyses when she writes that "textually-mediated discourse is a distinctive feature of contemporary society" which "must not be isolated from the practices in which they are embedded and which they organize" (1988: 38–9).

Because discourse and knowledge are constitutive of Man/Woman as subject, "identity" and "subjectivity" become dubious concepts under a Foucauldian gaze (Foucault 1972; 1979; 1980). Linda Alcoff speaks to this tension when she writes "In attempting to speak for women, feminism often seems to presuppose that it knows what women truly are, but such an assumption is foolhardy given that every source of knowledge about women has been contaminated with misogyny and sexism" (1988: 406). Similarly, and in earlier formulations, feminists of color have pointed out the danger of a feminism which ignores its own racial location (hooks 1984; Hull, Scott, and Smith 1982; Moraga and Anzaldua 1983; Smith 1983).

In return, this postmodern suspicion of subjectivity is problematic for feminist, anti-racist, and postcolonial thinkers especially when they are beginning "to remember their selves and to claim an agentic subjectivity available before only to a few privileged white men" (Flax 1990b: 220). In a context where politics finds more and more its point of departure in subordinated people's claim for their marginal identities (Melucci 1980; Wexler 1987), and their disidentification with hegemonic representation of them as others (Roman and Christan-Smith 1988), it is important to find ways to conceptualize women, blacks, Arabs, gays and other Others as subjects in their own right. Speaking more specifically about black experiences and subjectivities, bell hooks suggests however, that a critique of essentialism – when it is not solely the province of white intellectuals and when it critically engages with its exclusionary practices – can generate a liberatory and oppositional affirmation of "multiple black identities, varied black experience. It also challenges colonial imperialist paradigms of black identity which represent blackness one-dimensionally in ways that reinforce and sustain white supremacy" (1990: 28).

In interesting efforts to come to terms with this set of contradictions, de Lauretis and Alcoff argue that "identity" – a necessary component of the subject's agency – is attained through an ongoing process of people's self-analysis, interpretation and "reworking" of their actual social positions and meanings given to these positions through discourse (Alcoff

1988; de Lauretis 1984). In light of this brief theoretical discussion, and paraphrasing Linda Alcoff, I would define identity as the product of an individual or a group of individuals' interpretation and reconstruction of their personal history and particular social location, as mediated through the cultural and discursive context to which they have access. Similarly, I would define the process of identity formation as the set of self-definitions and practices through which people constantly modify this construction.

METHODS AND SOURCES OF DATA

This study is based on data collected during 1989–90, when I spent several days a week in the vocational Lycée Lurçat.[1] The school, located in a working-class suburb of Paris, offers primarily short clerical training programs to its students, most of whom are female. At the time of the study, approximately 70 percent of the students were female, half were of migrant descent or were migrants themselves, and over 25 percent were of North African parentage.

I "followed" two classes of girls in a short secretarial track, who were studying for a *Brevet d'Etudes Professionnelles* (BEP = Vocational Studies Certificate). Among the two classes (48 students), 20 students were French, 11 were of Algerian descent, and the rest were of other migrant parentage. Only one student was male. I spent the first trimester observing classroom processes and the last two trimesters doing semidirective interviews (individual and collective) of students as well as intermittent classroom observations. I also interviewed school officials and informally talked to teachers.

ALGERIAN MIGRANTS AND THE FRENCH SCHOOL

What I want to explore first is the specific positioning of the Algerian community in France *vis-à-vis* schooling and education and its particular manifestations where young women are concerned. It has been argued that schooling and education are of prime importance for a community whose collective trajectory is that of rupture and displacement. Migration itself marks the beginning of that trajectory and school education for one's children represents one of the powerful means to further this process. Dispossessed of their land, Algerian men were theoretically "free" to leave their country and try their luck in France. In that limited space of choice, some decided to migrate, to change, "to give a different meaning to their lives." At first, the idea of leaving was closely connected to the idea of *going back* "richer, freer" (Begag and Chaouite 1990: 37). However, most stayed and settled with their families.[2] Again, in that very thin slice of choice – staying rather than returning to a postcolonial shattered economy and tormented civil society – the education of the children in French schools would perhaps fulfill the initial dream of being "richer, freer," but in France rather than Algeria this time.[3]

Where girls are concerned, this collective positioning in favor of school and education furthers the dynamic of rupture and displacement because it necessitates the breaking of certain "traditional" rules – and the taking of certain risks. For instance, girls must be allowed to venture into public space. This breach in the "law" of the Father by fathers and brothers themselves is portentous of future trouble and has been beautifully captured (in the Algerian setting) by the Algerian writer Assia Djebar. Interestingly, Djebar draws links between knowledge, sexuality and self-realization for young Algerian women.

> As soon as the young girl goes out to learn the alphabet, neighbours cast the sly look of those who are sorry 10 or 15 years in advance, for the audacious father, the irresponsible brother. Doom is bound to fall on them. Each knowledgeable virgin will know how to write, will surely write "the" letter. The time will come for her when love that can be written is more dangerous than love that can be sequestered.
> (Djebar 1985: 11, translation mine).

Young women in the changing context of the Algerian community in France seem to be called to play contradictory roles: on the one hand they are still considered the bearers and keepers of "tradition" and culture, and on the other hand they are expected to engage with a new culture and new traditions.

In the interviews I conducted, mothers stand out as an important force around this issue. Even when not directly telling their daughters to stay in school, their experiences and stories strongly suggest the importance of school for their children, and for girls in particular.

> Aïcha: My mom never went to school. I get the feeling that she's telling us that we [the girls] should keep on studying. Since we have the opportunity, she wants us to stay in school; it's fundamental for me too. . . . Yes school is useful, it can help me enjoy all the things that my mother has missed; youth, work, *real life*. (emphasis added)

It's not only mothers who – through their very lived experiences – suggest the importance of school. Parents in general play a similar role and sometimes even a father will push for his daughter's education.

> Soraya: My parents want me to stay in school, they don't want me to get a bad job. . . . The job my father had at the plant was hard. He's still working very hard; he doesn't want me to get this kind of job. He wants me to get a good job, a good salary, and most of all a good education, because he never got the opportunity to study.

82

What I find interesting here is the fact that "real" material conditions – postcolonial economic relations between France and Algeria, the world economic crisis, French deindustrialization, and migratory flows – as well as the discursive insistence of the idea of school as opportunity, create specific boundaries within which young women of Algerian descent in France come to construct themselves within and in relation to school. The positioning in favor of school and education within the Algerian community in France, and its resonance among young women within that community must also be looked at in light of mass discourse in the dominant culture.

THE FRENCH MEDIA, GIRLS OF ALGERIAN DESCENT AND THE SCHOOL

In France, in the recent past and within the larger context of the rise of all religious fundamentalisms, the media has fuelled racism and xenophobia by constructing the "Arab" (in France read: the migrant) as a potential threat to national unity. According to these constructions, men are caught in "tradition" and "religious fundamentalism" and women silenced and oppressed by Islam.[4] Second-generation immigrants as they are often labeled (the term itself locks youth of migrant parentage in a fictive temporary state in the host country), are often presented as a "social time bomb."[5]

Young males, in particular, are often depicted as potential or actual delinquents. Girls of North African descent however – in part because of their supposedly better results in school – are increasingly represented as a potential integrating/assimilating force for the North African communities in France (Andre 1991). While doing my research at Lurçat, Paris billboards and press stands regularly presented headlines and images constructing this powerful scenario:[6]

Le Point, 6 March 1989: "France: the Islamic shock"
The picture shows Moslem men praying in the street during a post-Rushdie demonstration in Paris. The caption reads "France counts more Moslems than all the Arab Emirates."
L'Evénement du Jeudi, 6–12 July 1989:
"The courage of the *beurettes*': they are doing better [in school] than their brothers and they might be the bridge between two worlds"
Le Point, 7–13 August 1989: "Immigration: a bet and its risks"
The picture shows two North African men getting into a police car. One of them – young – is handcuffed, the second is gathering suitcases closely watched by a French policeman.
l'US (Teachers' union magazine), 8 December 1989:

"Islam, an obstacle [to integration/assimilation]?"
Le Point, 5 February 1990: "*Beurettes*: Integration in the feminine form"

As these captions and images suggest, while men are represented as criminals/religious fanatics – potential outcasts in any case, young women emerge as the "bridge between two worlds." The best example of this mediatic construction which developed in the fall of 1989 and lasted for most of the academic year 1989–90 is what has been called in France the "Veil Case." On 18 September 1989 a junior high school principal suspended three North African girls from a school in a Paris suburb because they refused to remove their *hijeb* (koranic scarf) in school. In a few weeks "the veil case" became a national debate dramatically splitting France into two opposing camps. One side favored the exclusion of religious signs in public schools and saw "the veil" as a symbol both of militant fundamentalism and of the oppression of women. The other side condemned the exclusion of the three girls on anti-racist and multi-cultural grounds and often argued that being in school would be the best way to "save" young women from Islamic fanaticism and gender oppression. Indeed, in the press, "the veil" (note the linguistic slippage from *hijeb* to "veil") was often conflated in a single symbol Arab, Woman, Islam and Tradition.

Conflicting discourses depicted the French school – and its secular tradition – as the target of Islamic militants in their holy war against the French nation (Le Pen's[8] scenario) or as the powerful tool for the assimilation of North African migrants' children (the liberal scenario). School in either case is central and female students of North African descent are center stage but are represented without a voice of their own; even when shown with a clenched fist it is assumed and/or suggested that they must be manipulated by their fathers and their brothers (*Le Nouvel Observateur*, 26 October–1 November 1989).

WORK OF IDENTITY FORMATION IN THE SCHOOL

When mass culture is constructing such powerful representations of a certain group of people, when so much has already been said and written about "them," the work of self-formation – the construction of a "we" – becomes very tricky and difficult. It is to this process of self-definition that I will now turn by analyzing the ways in which several girls at Lurçat understand and interpret their position as young women of Algerian parentage in French society. This work of self-definition necessitates a careful negotiation and recoding of the symbols and common sense ideas widely spread in mass culture. It also necessitates exchange and communication. School, as Djamila, one of the students, aptly points out, offers a space where this kind of exchange can happen for girls of Algerian descent.

Djamila: You can criticize school all you want, but at the end of the day that's where you learn a lot. Okay, there's also TV, but with your parents . . . even among the French they don't talk about certain things like menstruation, drugs etc. Okay, I'm not saying that teachers talk easily about all this but at school you can know more and the contact with your schoolmates teaches you a lot.

The learning and knowing to which Djamila seems to be referring also happens in collective subjective positioning, in the difficult game of defining "us" *vis-à-vis* "them." In the two classes I observed, ethnicity plays a central role in the ways in which individuals and groups position themselves. In both classes, young women formed friendship networks primarily within their own ethnic circle; girls of French descent sat with and hung out with other French girls. Similar patterns were found among girls of Portuguese descent and girls of Algerian descent.

For girls of Algerian descent, the locus of difference between themselves and their schoolmates of French descent lies in their respective relation to sexuality. In an informal interview for instance, Soraya, Acia and Farida brought to the fore the issue of sexuality which they perceived as the main difference between them and the French students and which made it very difficult for them to feel comfortable among French girls.

Soraya: All they talk about is sex, sex, sex! After a while we get fed up, it's too much in the end.

Acia: That's right, that's all they have on their mind: sex!

Farida: It's true, we're more comfortable among ourselves.

Acia: Yeah! See, we've been in the same class for almost two years but the other day I got here late and there were only French girls and I felt uneasy.

In an individual interview another student added:

Laïla: They are freer. Their parents give them more freedom. They can do what they want. . . . Some of them take advantage of this freedom in ways that I really dislike. They dress crazy, they sleep with several guys, I think it's . . . stupid. Anyway I'm of Algerian descent, I don't envy them too

85

> much. I feel that because of all this freedom they have they go overboard, it's getting ridiculous.

When asked why she thought the two groups of young women had such a different attitude toward sexuality one student answered:

Soraya: It's French society, because now French society does not keep its customs, see. For us our culture [is important] . . . but for the French, no, they don't keep their culture. . . . When they are told you should not do this, the Bible says this or that, they don't care, it surprises us.

It is interesting to note that Soraya conflates customs, culture and religion. We get the feeling that girls of Algerian descent know that French girls have a greater latitude in general and particularly in relation to sexuality. However, none of them expressed the desire to enjoy similar freedom. This might be simply a realistic appraisal of what is possible in their families and their communities, or a refusal to totally disidentify with Algerian customs, culture and religion. But it might also reflect a subtle understanding of the pernicious side effects of pseudo sexual freedom for women in a society shaped by otherwise deeply entrenched gender inequalities.

The particular ways in which girls of Algerian descent understand the issue of sexuality, and how they position themselves in relation to it, needs to be explored from various angles. While they seem to take on a rather "conformist" stance when comparing themselves to girls of French descent, they nevertheless take on a critical point of view *vis-à-vis* their own community and its particular gendered boundaries.

Indeed, the interviews suggest a great awareness of gender asymmetry within the family among girls of Algerian parentage. This awareness seems to be linked not only to actual differential treatments of male children within Algerian households,[9] but also to the fact that the issue of gender oppression in Islamic communities is part and parcel of French mass discourse on immigration.

The following excerpts discuss whether my informants felt their parents treated them differently from their brothers.

Fella: For us, the girls, my parents like to know who we go out with. It's true, they kind of like to know. . . if I tell them at what time I'll be back they will always worry more than for my brothers, see . . . Okay, it's not like I am locked up or anything but with my parents it's always "who's that boy? bla bla bla" – see what I mean? Things like that. But, it's alright, I'm not locked up – it's okay because sometimes I can go out with my brothers.

Aïcha: My parents want the same thing for their sons and daughters as far as school is concerned, but in the family it's a different story. Boys have more freedom. They can go out at night. . . . My parents don't trust us (the girls). If we go out with male friends; we must be doing . . . dishonest things (she laughs). They think we are wild.

These responses clearly underscore that girls of Algerian descent feel much more controlled in their movements than their brothers. On the one hand they resent the existence of such control. Aïcha, for instance, using moderate language, laments that "Boys have more freedom," "My parents don't trust us." She also hints at the driving force behind such control: "When we go out with boys we must be doing dishonest things." As I will show later, those young women are quite aware that their use of space is tightly controlled because their parents, and their fathers in particular, want to keep a check on their sexuality.

On the other hand, they seem to accept, at a certain level, the legitimacy of such constraints. Fella's insistence on the fact that she's not "locked up" can be interpreted here as an effort to dispel the idea that Algerian girls are ultimately without freedom and agency – an idea widely spread within the dominant French discourse.

These responses begin to illustrate the complex process of identity formation in which girls of Algerian descent work/play to self-(de)construct themselves (de Lauretis 1990: 136). What we have here is neither pure resistance nor pure accommodation to already available scripts (i.e. the liberated French girl/the traditional Algerian girl). What seems to emerge echoes Martin and Mohanty's reading of Minnie Bruce Pratt's attempts at self-definition as involving "a series of successive displacements from which each configuration of identity is examined in its contradiction and deconstructed but not simply discarded" (de Lauretis 1990: 136).

One of these configurations lies in the critical gaze of these young women on the gendered organization of their families. Indeed, in the interviews, girls of Algerian descent point out that boys get more attention and greater care from Algerian parents. Nasma's comments are quite telling and stress her own discontent at such blatant preferential treatment.

Nasma: My mom, she adores my brothers! She's always after them. She talks to them like they are babies "Have you eaten? You haven't eaten, have you? Should I give you something to eat?" But, us [the

girls], we could be dying right there, she wouldn't
lift a finger!

Boys have certain advantages, certain privileges because they are
thought of by Algerian parents as the ones who will be in charge of
the family. Acia told me that her parents would prefer it if she quit school
after this cycle of study, but when asked if they would react the same way
if she were a boy, she answered:

Acia:	No, if I was a boy they would prefer me to stay in school. Because when a boy gets married he must be in charge; he's the head of the family, he must . . . I don't know. But for a woman it's different – if she doesn't find a job, she can always stay at home.

Anifa elaborated further what was expected of boys in the Algerian
family and how that affected the ways women were treated and thought
of by their parents.

Anifa:	If there is a woman in the family they are more careful with her. They go "he's a boy, he's *the man*." And he can go out too, he doesn't even have to make his bed! . . .
C.R.:	The man?
Anifa:	Yeah, a man, he's the man in the family. He's the one in charge . . . I can't stand it! They [Algerian men] are very proud, I mean. A Man . . . my father tells me, he makes me feel it . . . a woman is nothing . . . he really makes me feel it and that's why I want to show him the opposite. See, in fact I'm struggling for that too, to make him understand that perhaps I can *make something of myself.* I hope so. (emphasis added)

Anifa, who openly resents this unequal treatment, sees in her education (and in staying in school) the possibility of challenging what her
father and the men in her community think of women and of transcend-
ing the future they have in mind for her. The unequal treatment sensed
by girls of Algerian descent also gets translated in the ways they describe
their brothers taking on the role of male controllers in the family and the
community. Soraya broaches this issue when talking about her rapport
with her male siblings:

Soraya:	I personally get along with all my brothers. But my brother always tells my sister what to do "You go out too much! You're wearing too much make up!

You think you live in a hotel, or what?" And yet
she is the oldest!

Soraya, who has chosen not to go out and to concentrate on her
education, does not encounter the rebuke of her brothers. Her sister,
on the contrary "goes out a lot, is always on the phone, complains all the
time and fights against [their] parents." Such assertive behavior, however, is met by her younger brother's disapproval. Soraya's sister is
seeking self-determination outside acceptable boundaries and that gets
her into trouble with the male members of her family. Soraya however,
never clearly expressed her resentment at this situation even when I
asked her directly.

Soraya:	With the Moslems it's like that, you know. If we want to go out, if we want to do what we haven't been able to do when we were younger, we must get married. If your husband does not let you go out, then it's too bad . . .
C.R.:	Does it bother you?
Soraya:	Frankly no! It's never bothered me. It doesn't bother me at all.

The potential contradictions in Soraya's interviews are also a sign of
the lack of "purity" within these emerging identities. This lack of purity
reminds us that identity "is a locus of multiple and variable positions"
which at times may be "ideologically with the 'oppressor' whose position
it may occupy in certain socio-sexual relations [if not in others]" (de
Lauretis 1990: 137).

In another set of interviews, these young women, individually and
collectively, critically analyze how particular social constraints work to
define the boundaries of acceptable female behaviors particularly in
relation to sexuality.

Acia:	I can receive as many phone calls as I want to, as long as they are not from men! (They all start laughing)
Farida and Soraya:	Same here!
C.R.:	Really?
Soraya:	Of course, no phone calls from boys! My mother never stops telling me: "Beware of boys!"
Acia:	Sure! My mom, ever since I turned eight, she's been telling me "Watch out for the boys!"
C.R.:	What about you Farida?
Farida:	They don't say it, but I can guess it. They make me understand. They don't gossip or anything, but I can hear when they talk about certain girls. So I

figure that they'd feel the same way about their daughter!

These comments reveal a strict check on female sexuality in their respective families. While none of them is willing to criticize this control in front of an outsider (a potential arrogant eye), their comments and their insiders' laughter might suggest it is something they talk and perhaps complain about among themselves.

In the same vein, Anifa, one of the most rebellious of the North African girls, explicitly points out that it is indeed her sexuality that men, and her father, in particular, are trying to control:

> *Anifa:* Sometimes my father sees me and I'm talking with a guy and I don't know but deep inside it must break his heart to see me. For him it's like, I don't know, to see his daughter talk with a boy. I try to tell him "It's nothing; he was just asking for directions" or something, "that's all." But he cannot accept it, I don't know, and it also has to do with being a virgin and the whole deal, see. They smother me – "don't do this, don't do that" – because I am a girl.

Abla, who is less openly rebellious than Anifa, also clearly underlines the link between young women's inability to move around freely and male control over their sexuality.

> *Abla:* Being a virgin is really important for us. For our parents it's like a matter of honor. That's why we cannot go places. But for a boy it doesn't matter; it bothers me that they make a difference between boys and girls. My mother does not make a difference but my father does; he's old-fashioned, backward.

Of all these students Djamila is the most articulate about and the most angry at such inequalities.

> *Djamila:* It's because at home he's [the Moslem man] got all the power. He can do anything! He can shut you up! He can tell you "do this, do that," we can't say a thing! Even with my brothers! Yeah, my brothers, for instance, with housework. These are little things but they kill you, they kill the life of a person. They don't do a thing. . . . Even when they are born in France, you get the impression that they think like men who are born over there

[in Algeria . . .] Because we are girls we cannot do anything! As women. Because we've got a different sex than they have, that's it. I can't stand this way of thinking! . . . Men have got everything; when they come home, perhaps a couple of them clean up. But men are waited on, whereas the woman does her schoolwork and after that she's got to prepare dinner, she's got to do house chores. She tries to do it all because she tells herself if I don't do it, if I don't go to school I know I'll get married at fifteen, at eighteen.

Djamila, like Anifa, stresses the importance of school for North African girls. At home, some of them "can't say a thing" and school in this context represents a space where women can escape to a certain degree the male gaze of family members:

Anifa: I told my father that he is not always behind my back, I tell him that at school I do talk with boys and that it doesn't mean that I'm getting into trouble, and all that. When I talk like this he gets really quiet but it breaks his heart because I'm telling him the truth, see.

CONCLUDING REMARKS

School, then, emerges as a space where young North African women can bide time, where perhaps they can "make something of themselves," where, in spite of the double burden described by Djamila, they gain some control over their lives. While school is often presented in liberal discourses on immigration as a great "assimilating" tool, it is clear that it does not necessarily "erase" differences and turn children of migrants into bona fide – even if second rate – French products. In fact, as we have seen, female students of Algerian descent are actively engaged in the work of positioning themselves between, within and against different discourses, and school, in this respect, offers them a space to conduct this work.

The particular positioning of working-class female students of Algerian descent in a French vocational school in the late 1980s puts them at the crossroad of several contradictory discourses. Strong media projections of them as at once oppressed within their own communities and as the potential integrating agents of that community into larger French society; mounting xenophobia and racism; and conflicting messages they receive from their own families and communities regarding school as an

91

appropriate space for them and as a step toward advancement with accompanying clear and restricting delineations of their roles within the community and society at large.

This specific location – which starts, as discussed at the beginning of the paper, with immigration as a trajectory of displacement – seems to create for them the possibility, in certain circumstances, of critical insights (of multiple displacements, dislocations) and the potential for a rebellious consciousness. At the same time it can also anchor them in some of the more "traditional" values of their communities. To resolve and envision the potential outcomes of these contradictions is definitely beyond the scope of this paper. However, this close "micro" analysis of identity formation might enable us – in our empirical works – to explore beyond the limiting dualities of resistance/accommodation, freedom/determination, and to conceptualize identities as multiple, varied and constituted through an ongoing struggle to re-invent selves within specific material and discursive boundaries.[10] Furthermore, it suggests the importance of moving beyond representations of a homogenous, Euro-American notion of Northern womanhood, to an acknowledgment that the North as well as the South is a site of difference and diversity among women.

NOTES

1 For a full discussion of the study see Catherine Raissiguier (1994), *Becoming Women/Becoming Workers: Identity Formation in a French Vocational School*, Albany: State University of New York Press.

2 Recent French immigration policies have favored immigration on a permanent basis for migrants of European origin but encouraged returning home for African and North African workers (Abadan-Unat 1984). In 1974 the French government halted all work migration and allowed migrants to enter and remain in the country for family reunification purposes only. Ironically, the policy changed what had started as a transient male worker immigration into an immigration of settlement.

3 This relation of trust and hope *vis-à-vis* the French educational system is far from monolithic. Some Algerian families perceive and resent the French school as one of the many barriers which prevent them from achieving social mobility (Zeroulou 1987; 1988). In spite of a remarkable broadening of the social base of the French educational system since World War II, educational outcomes are still widely unequal (Baudelot and Establet 1971; Bourdieu and Passeron 1964). Working-class youth and students of migrant parentage are systematically overrepresented in non-prestigious short vocational tracks (Boulot and Boyzon-Fradet 1988).

4 This is also happening in "high" academic discourse. For a critique of Eurocentric analyses of Islam and of women in Islamic societies see Mernissi 1975; Ahmed 1982; and particularly Lazreg 1988.

5 For a discussion of the phrase and its implications in the former West Germany in particular, see Castles (1980) and Castles and Kosack (1985).

6 The papers and magazines selected here cover a wide spectrum of political agendas and all belong, except for the *US*, to the non-specialized media.

7 *Beurette* is the feminine form of *beur* which is a slang word used to describe youth of North African parentage in France. While initially the term was coined and used by the youth themselves, it has now been recuperated by the dominant culture. It is interesting to note that *beur* is a double reversal of the original term *Arabe*. French working-class youth talk *verlan* (reversed talk); in this game of language *Arabe* became *rebeu*, and then again *beur* (Aissou 1987).

8 Jean-Marie Le Pen is the leader of the National Front (FN) in France which has been at the forefront of all racist, anti-immigrant campaigns of the past ten years. This political party, which was a marginal force in the 1970s, now carries more weight than the Communist Party (traditionally the fourth force in the French political scene).

9 Of course asymmetric treatments of boys and girls exist within French families also, but the locus of the difference is less visible perhaps because it complies more with wider societal gender practices.

10 For recent discussions and explorations of these possibilities see Kondo 1990; Kruks 1992; Mahoney and Yngvesson 1992.

5

GENDER, HISTORY AND IMPERIALISM

Anne Wheeler's *Loyalties*

Julia V. Emberley[1]

[T]he relationship between global capitalism (exploitation in economics) and nation-state alliances (domination in geopolitics) is so
macrological that it cannot account for the micrological texture of
power. To move toward such an accounting one must move toward
theories of ideology – of subject formations that micrologically and
often erratically operate the interests that congeal the macrologies.
Such theories cannot afford to overlook the category of representation in its two senses. They must note how the staging of the world in
representation – its scene of writing its *Darstellung* – dissimulates the
choice of and need for "heroes," paternal proxies, agents of power –
Vertretung.

> G.C. Spivak, "Can the Subaltern Speak?" (1988: 279)

INTRODUCTION: WE WON'T PLAY BODY TO YOUR TERRITORY

In Barbara Kruger's signature photomontage, a photographic image of a
woman's face with each eye covered by a solitary leaf is cut across the
bias with the following text: "We won't play nature to your culture."
Kruger's image both visually and discursively deconstructs the hierarchical positioning that structures the nature/culture duality. As Craig
Owens elaborates, this image presupposes "a binary logic of opposition
and exclusion that divides the social body into two unequal halves, in
order to subject one to the other" (Owens 1983: 5) The order of
subjection mediating Kruger's graphic display is specifically gendered
by the representation of a naturalized female figure which Kruger's text
equally graphically sets out to denaturalize. This denaturalization of the
female figure enters into the deformation of Kruger's text that appears in
my title to this introduction, "we will not play body to your territory," so
as to recall a prevalent trope currently mediating postcolonial discourses:

the feminization of colonial territory. In its metaphorical expression this trope figures the Woman/Body as a landscape of natural desire, the shape of imperial expansion; metonymically, the Woman/Body, in its most fractured and reified form, is disassembled into a synecdochic machine reproducing an originary fantasy of the virginal site that entices imperial penetration and conquest. This figural investment in an unspecified and seemingly generic Woman/Body for First Nations' women, whose bodies are, in reality, subjected to colonial domination, represents an important area of study for decolonialist and materialist feminist critical practices. .

Anne Wheeler's film *Loyalties* provides an occasion to illustrate some of the contradictions which emerge when feminism takes up the trope of the feminizing of colonial territory, not necessarily to de-feminize and hence denaturalize its metaphorics and metonymics, but rather to deploy this trope in order to critique a patriarchal narrative of imperial history.

THE LOYAL OPPOSITION

Loyalties was released in Canada in 1984 to very little critical attention. The importance of the film lies in its staging of the tensions and contradictions between Native and Anglo-Canadian women, not unrelated to similar issues being raised in the political arena at that time. Set in a northern Albertan town called Lac LaBiche, *Loyalties* is a story about a relationship between two mature women: Rosie Ladouceur, a Native woman, possibly Cree or Métis, who lives with her mother and children on a nearby reserve, and Lily Sutton, an English, bourgeois woman, who has recently immigrated from England. What Lily and Rosie's names connote in terms of flowers and colours (note too their syntactic ordering here) – white and red, the flesh and blood, inside/outside duality of Christian and racialist ideologies – already signals the naturalized, oppositional subject positions they will occupy in the film. Their relationship follows a series of twists and turns, involving class antagonism, mutual friendship, intense hatred and, in keeping with the film's melodramatic narrative structure, a final reconciliation. Their friendship dissolves into a form of absolute hostility when, in the climactic scene, Lily's husband, David Sutton, rapes Rosie's adolescent daughter, Leona. At the conclusion of the film, Rosie and Lily consolidate a necessary alliance in their respective struggles against a common enemy: David Sutton, who figures as the patriarchal embodiment of British imperialism.

Loyalties is a film that in no easily or immediately recognizable form is about the history of British imperialism. The film neither rewrites the historical narrative, filling in its homogeneous empty time by retelling the "story" as it were, nor does it simply reduce the status of that history to an atemporal flow of random or aleatory events. Imperial history is radically decontextualized from its normative historical continuum and re-situated in a spatial, in-temporal dimension, something approximating Walter Benjamin's *Jetztzeit* (now-time). Within the realm of the "now-time," repeatable moments stretch like dots across the diachronic axis of developmental imperial historiographies. As an example of a repeatable moment, the rape scene at the film's conclusion affords, in its synchronic structure, a view into the inner text of colonial violence.

The inner text of colonial violence can be read in *Loyalties* as a gendering of British imperial history through the text of sexual difference, a text on which and through which is written the fractured identity of the Native/Woman/Body as a bearer, rather than as an agent, of imperial significations. Rape comes to represent the event through which the text of sexual difference signifies. Thus, in the sexual/textual violence of colonial discourse, the visual language of sexual difference maps onto the body of a Native woman and imperial history itself is cognitively mapped on that same body. As Jenny Sharpe observes in her discussion of the sexual discourse of rape in E.M. Forster's *A Passage to India*: "Since it articulates the contradictions of gender and race *within the signifying system of colonialism*, the sexual discourse of rape is overdetermined by colonial relations of force and exploitation" (Sharpe 1991: 36). In other words, as the body of an indigenous woman speaks the violence of colonialism, that site of enunciation (*parole*) finds its signifying system within the discourse (*langue*) of colonialism. Native and non-Native women's bodies in Wheeler's film, their figuration, characterization and relationship to each other, "speak" the violent disruption of colonial relations. Wheeler's use of rape, however, problematizes this inscription of colonial violence by situating it exclusively in terms of patriarchal domination; hence her visual (re)presentation of rape produces an ideological effect in support of a feminist allegory of global sisterhood for which rape, as a mode of explicitly sexualized violence, constitutes a transcultural and hence unifying experience of oppression.

Louis Althusser's re-writing of the concept of ideology as an imaginary resolution on the part of individual subjects to the contradictions of their real lived experience is indispensable to an analysis of ideological effects produced by this liberal feminist ideal of global solidarity (Althusser 1971: 162). Using Althusser's definition of ideology makes it possible to "read" *Loyalties* as an aesthetic mode of representation which attempts to resolve the political contradictions of First Nations/white-colonial women's solidarity for the women's movement in Canada during the early 1980s. *Loyalties* imaginatively expresses the desire for solidarity on the part of a predominantly white, middle-class, women's movement. As I discuss below, the film's diegesis and Wheeler's use of a parallel editing

96

technique produce a story that resolves, that is, covers over, the contradictions on which that wish-fulfillment for solidarity rests.

Loyalties exemplifies a feminized version of colonial history at the same time as it masks a critique of the totalizing tendencies of a feminist rewriting of the history of colonial expansion and domination. In order to elaborate the contradictions the film generates, it is important, initially, to situate *Loyalties* in the context of a political movement under way during the time the film was made and released: the political struggle Native women were fighting with the Canadian state to remove those sections of the 1876 Indian Act which specifically discriminated against Native women.

THE POLITICAL HORIZON

During the late 1970s, the Tobique women of Eastern Canada lobbied the Canadian government to amend sections of the 1876 Indian Act which specifically discriminated against Native women. The clause under scrutiny stripped Native women of their status or band affiliation if those women married non-status Native men or non-Native men. Such a policy clearly aided colonial objectives to assimilate Native peoples in two ways: by reducing the numbers of Native people with a claim to status and band affiliation (the children of these women also lost their status), and by attempting to impose European patriarchal norms on gatherer/hunter cultures most of which were characterized by egalitarian gender relations. When the Canadian judiciary system failed to recognize the validity of Native women's charges of sex discrimination in the Levall–Bedard case, discussed in more detail below, the Tobique women took their complaint through the Sandra Lovelace case to the United Nations and won their cause to the embarrassment of Canada's legal institutions and civil liberties record. The United Nations' condemnation of Canada no doubt provided some ethical support for the eventual passing of Bill C-31. However, it was largely the Charter of Rights and Freedoms in 1982 (in effect in April of 1985) that created the legal basis for challenging the Indian Act.

The passing of Bill C-31 in June of 1985 was acclaimed by the women's movement as a significant victory in the struggle against patriarchy.[2] Patriarchy can be understood in this context as an ideological enclosure, strategically containing indigenous cultures within the limits of a European epistemology.[3] However, this construction ignored the fact that patriarchy has been experienced differently by Native and Euro-Canadian women. Whereas Euro-Canadian women have been seen as the bearers of a nation-building enterprise, Native women have been perceived by the state as barriers to achieving national sovereignty.

In the aestheticization of Native women's contemporary experience, the Native/Woman/Body comes to articulate a new hybrid signification (while still retaining the layers of residual meanings acquired in the late nineteenth century), a mixture of historically and culturally discontinuous meanings, referents and values. On the one hand, this new hybrid figure aligns First Nations' territorial land claims with a feminist critique of the ideology of sexual difference: land and body become marked as one and the Native/Woman/Body figures as a utopian site upon which to inscribe the dream of an organic community. On the other hand, this set of significations further encodes a natural, holistic trinity, Native/Woman/Body: the feminization of Native culture, the feminization of woman, the feminization of the body: objects of conquest, penetration, territorialization and control. Meanings proliferate exponentially, compounded by violence, naming, language. As I will elaborate in the following section, the narrative structure of *Loyalties* constitutes both a thematic and historical documentation of this aestheticization process in which the political realities of Native women are circumscribed by the epistemic imperatives of feminist postmodernity.

THE AESTHETIC HORIZON

The narrative logic of *Loyalties* unfolds through a parallel editing technique. Through juxtaposed *mises en scène* – contrasting scenes between Lily Sutton's newly built, expensive log house and Rosie Ladouceur's farmland cabin, perhaps situated on an unspecified reserve – a sharp distinction is constructed between two cultural and class-specific domestic spheres. Brenda Longfellow discusses *Loyalties* in the context of the "representational contradictions of melodrama" (Longfellow 1992: 52) the film both appropriates and exploits. She notes the movement of parallel editing in the construction of differences between the two women:

> The counterpun[c]tal rhythm is repeated throughout the film in a series of paired sequences. Different attitudes towards sexuality are dealt with in the paired sequences which first reveal the frustrations of Lily and David, which is subsequently contrasted with the sexual reconciliation of Roseanne and her estranged husband, Billy, which is tender and toughly passionate.
>
> (Longfellow 1992: 53)

Longfellow concludes that these contrasts are eventually "soldered through another kind of temporality which works toward, not a suspension of these differences between the two women . . . but toward mutual recognition and solidarity" (Longfellow 1992: 53–4). However, I argue that, whereas the notion of parallel editing focuses on the equitable distribution of gendered cultural differences, thereby conforming to the film's structurally gendered logic, the concept fails to address just why the gendering of private/public, domestic/foreign, inside/outside dualities are necessary to the very construction of the sexual differences the film sets out to resolve. The suspension of differences in the move toward "mutual recognition and solidarity" is precisely the ideological problem the film poses, and, what needs most urgently to be addressed. While the relationship between a Native woman and a bourgeois, English woman takes its narrative construction from the film's parallel editing technique, it is the visual language of sexual difference that performs a central role as the structural mediation through which the gap left by the parallel editing technique is temporarily sutured.

Some examples from the film will help to demonstrate the discontinuities generated by the film's structural effects:

In a set of introductory scenes, Wheeler produces an oppositional spacing of gendered class-divisions: Lily Sutton is shown meeting her husband, David Sutton, a medical doctor, at the airport, having just arrived from England to settle with him in Lac LaBiche. While they sit in a hotel restaurant with two of their children and two friends of David's, eating and talking, the scene cuts to a bar-room in the same hotel, where Rosie is waitressing. A verbal fight between Rosie and her boyfriend Eddy results in Eddy punching Rosie in the face and knocking her to the ground. In the next scene, the hotel manager comes up to David Sutton and asks him to attend to Rosie's facial injury. David dutifully goes off to examine Rosie and takes her to the medical clinic to administer stitches. Though these scenes edit back and forth between Rosie/Eddy and Lily/David, they take place within the same spatial dimension, the hotel. The spatial continuity between these otherwise opposed scenes provides a rationale for their point of intersection. It is, in part, this loss of spatial unity or the introduction of spatial discontinuity in the concluding violent rape scene which will contribute to the sense of anxiety and panic Wheeler is able to produce over who belongs where and in what context. In other words, the pastiche-like approach to framing spatial discontinuity in this parallel editing technique itself represents a process of physical and psychological territorialization, a re-mapping of the body of woman on to the colonial parameters of territorial conquest.

Three other scenes, also early in the film, construct the spatial discontinuities in Lily and Rosie's cultural frames of difference. In the first scene Lily reads a bedtime story to two of her children. Along with David Sutton, who is eavesdropping on the bedtime reading, "we," the audience, also overhear the following story by Lily:

In times of yore, when wishes were both heard and granted, lived a king whose daughter was so beautiful that everyone in the kingdom thought she must be very special. They worshipped her and said she was the most blessed person on earth. In truth the princess was a very lonely child and she had but one wish in her heart. She wanted a friend. Someone who wouldn't judge her by her beauty and her wealth. To see into her face, she looked so sad.

The sad story of the motherless princess virtually describes Lily Sutton's life. An only child, raised by wealthy parents, whom she rarely saw because of a boarding school education, Lily's relationship to her mother is a distant and emotionally unsatisfactory one. This distance was exacerbated by her marriage to David Sutton, who, although a professional doctor, comes from a working-class background. In a parodic imitation of her mother's class propriety, Lily tells Rosie: "Oh, one doesn't marry down. He's not our sort. His father works with his hands for goodness sake." Doubly exiled from family and country, Lily represents a kind of inverted female Crusoe, stranded on an island of material excess, yet deprived of meaningful social relations.

The scene following the bedtime story takes place at Rosie's mother's house. In this scene Rosie and her companion Eddy stand outside the house, embroiled in a verbal fight. During the course of their loud and raucous domestic squabble, Rosie refuses to accept Eddy's apology for his violent behavior the night before; the source of Eddy's violence apparently stemming from his jealousy and inability to accept that Rosie works in a bar where he likes to spend most of his time playing pool. As the fight escalates, Rosie criticizes Eddy for not being able to hold a steady job. Although he has held part-time work fighting fires, for Rosie, it's simply not enough. When Eddy threatens to hit Rosie for this insult to his masculine identity, Rosie's mother intervenes. She quietly tells Eddy to leave and urges Rosie to come indoors. While David Sutton observes Lily and the children in the previous set of scenes, during Rosie and Eddy's argument, it is their son Jesse and his grandmother who are shown watching through the cabin window.

While Lily Sutton may have material wealth, speak in the discourse of the proper and the civilized, have four children by the same man to whom she is legally and properly married, Rosie's unemployment (after Eddy punched her in the face in the bar, Rosie lost her job), her several children (not all of whom it is suggested Eddy fathered), her violent language, and her physically violent existence, do not deprive her of a mother to whom she can turn in time of need; the mother, security and protection, Lily Sutton's bourgeois existence would seem to prohibit.[4]

The third scene returns to Lily and David standing outdoors on the balcony of their new house. During the course of their dialogue, no violent language is exchanged between them, however, a sense of underlying tension is portrayed through David's distracted fumbling with a pair of binoculars and Lily's polite conversation. Their conversation reaches a threshold when Lily describes Lac LaBiche as forlorn and far away from everything. David responds by turning his binoculars toward her. Although Lily is standing no more than five feet away from him, he glances through the binoculars in a playful gesture and says "You look tired." Lily laughs and responds that she is tired, punctuating the remark with her own version of subdued playfulness. She calls him, in a soft tone of voice, a bastard. All joking aside, the mood is one of despair heavily masked by a coded bourgeois language and the controlled gaze of David's binoculars.

During the course of this dialogue, David expresses concern for whether Lily will get to like it in Lac LaBiche: "I suppose it is rather on the lip of civilization." He goes on to say, "We must look on it as an adventure." David's phrase, "the lip of civilization," suggests a telling metaphor of an anatomical deformation; a single lip of civilization, a single and unified image of the spoken word. The one lip is the sutured lip of Lily's silence about her husband's previous history of violent, sexual crimes in England – the reason for their immigration to Canada – and it is also a metonymic association of the virgin lips of Leona's body. David says "We must look on it as an adventure." The adventure of colonial conquest, the adventure of sexual conquest: this double writing of sexual and colonial conquest is produced upon a territorialization of the Native female body: as such, this body becomes a central figure in a postmodern battleground of textual and visual ideological struggles.

When Lily goes into the house to prepare dinner at the end of this particular scene, David looks out over the balcony through his binoculars. Eventually his eyes come to rest on Rosie and her children walking across the road to a path in the bush. In a self-reflexive gesture, in which the technological filmic apparatus calls attention to itself as the normal bearer of a masculinized "sadistic voyeurism" (Mulvey 1988), the camera reproduces the view from the binoculars, and the spectator's gaze, like that of David's is directed specifically toward the eldest daughter, Leona. Like a moving camera tracking an unsuspecting subject, David's gaze through the binoculars and that of the audience's pursues the image of Leona as she walks through the bush.

The binoculars, a binary set of lenses which the eye shapes into a unified field of vision, are an excellent metaphor for visualizing how the opposing domestic spheres created by the parallel editing technique are brought together by the visual language of sexual difference in the rape scene. While Lily laughs at David's playful distortion of her image in the binoculars, Leona, Rosie's eldest daughter, does not know that she is being watched; to paraphrase the text from one of Barbara Kruger's photomontages, David's gaze is hitting the side of her face (Kruger 1984: 9). Two separate women, two symmetrically placed lenses, two separate fields of vision, intersecting at a singular moment of focus; in this scene the focus of the imperial, phallocentric gaze is intently directed toward the young woman, Leona, who will be the victim of David's rape. It is Leona's body that will be the object and the visual spectacle of the violent collision of two worlds. The sadistic voyeurism of sexual violence is, itself, a visual analogue to the colonial surveillance of Native cultures during the process of cultural imperialism. Similar to the structure of epistemic violence, Spivak elaborates, that "constituted/effaced a subject that was obliged to cathect (occupy in response to a desire) the space of the Imperialists' self-consolidating other" (Spivak 1987: 209), the gaze through the binoculars creates a unified field of image-making; itself a gendered form of epistemic violence committed against the feminized heterogeneity First Nations represent to the colonial state.

Loyalties has more than one beginning: a scene appears at the start of the film decontextualized from the narrative continuity of the main story. Someone, dressed in rain gear, stands on the outside of a house looking through a window. We hear the rain and we hear a struggle taking place inside the house. We are not privy to the actual scene of struggle, nor do we ever see the face of the person gazing through the window. What is shown, instead, is a glass covered photograph of a woman with her husband and children shattered during the struggle. The film then begins again, as it were, with introductory credits and the beginning of the narrative proper. In retrospect, it would appear that this scene dramatizes an earlier rape by David Sutton witnessed by the eldest of David and Lily's children, Robert, who is apparently the person in the yellow raincoat. The rape occurred in England and the damaging publicity, it is suggested, is the reason for the Suttons' immigration to Canada.

This decontextualized beginning foreshadows the violence that is to come. And it does so by self-consciously reflecting on the limits and limitedness of various points of view: who is watching and what will be seen? Indeed, the opening scene is sufficiently ambiguous so as to suggest a series of potential spectators. While Robert could be said to be the most literal referent to this representation of the viewer, if not a voyeur, the sign of the anonymous spectator also functions as a catachresis, in the sense that the image of the spectator has no literal referent to which it can be adequately and fully fixed. Is this spectator Robert? or an omniscient figure such as the director, herself? or a mirror image of the audience? – a *mise en abîme* series of spectacular references which include a voyeur, a director and a witness. Wheeler could be said to be questioning the voyeuristic assumptions on the part of the spectator as well as her own role as director. But what she makes ambiguous in this primal scene she will concretize, visually, in the graphic depiction of Leona's rape by David Sutton.

The scenes leading up to the rape are also structured discontinuously between Lily and Rosie's celebration of Lily's birthday at a local tavern and Leona babysitting at Lily and David's house. Between the bar-room scenes and shots of Lily and Rosie driving home together laughing and singing, David is shown arriving home early from a fishing trip. He finds Leona babysitting alone in their house, clumsily attempts to seduce her and eventually rapes her. Lily and Rosie arrive to find Leona outside lying in the mud. When Rosie takes Leona away she grabs Lily's hair and says "What kind of a woman are you?" "Kind," here, is the operative word; for not only does "kind" designate gendered and racialized systems of differentiation that in *Loyalties* structure the asymmetrical relationship between the two women, Rosie's question also poses an ethical challenge to Lily to recognize the kinds of choices she is making both in her relationship to a rapist, her husband David, and to her friend Rosie, a Native woman. An acknowledgement on the part of Lily of the ethical asymmetries that condition these couplings would no longer deny the existence of difference between the two women, Rosie's otherness for Lily, and David's domination of Lily and Leona. Such recognition constitutes a mode of feminist justice, so eloquently summarized in the following passage by Drucilla Cornell: "For Levinas, to try to know the Other is itself unethical, because to do so would be to deny her difference and her otherness. Instead our responsibility is to hear her call, which demands that we address her, and seek redress from the wrongs done to her. The Other, then is 'there' in the ethical relationship, only as the subject's responsibility to her" (Cornell 1992: 88).

Althusser's notion of ideology as an imaginary resolution to the realities of social contradictions applies well to Wheeler's gendered narrative of British imperialist violence. If, for a moment, we situate those "realities" in terms of the gendered and racialized power relations that constitute Native/non-Native women's "friendship," and speak in the language of phallocracy, then we can speak about the phallus as the dominant sign and instrument of power around which and through which this intra-racial female relationship is consolidated. An instrument of counterinsurgency in this context (an instrument through which we can counter the loss of women's agency due, supposedly, to being phallus-less) can be found in Spivak's configuration of the womb as a "tangible place of production" (Spivak 1987: 79). We should also note here that this rape represents a brutal moment of the denial of Native women as agents of signification. By extending the discourse of phallocentrism, in which the phallus functions as an instrument of power controlling the constitution of transcultural and cross-racial female relations, to an analysis of the post-rape scene the following reading can be produced:

Following the rape, David Sutton sits in a chair in the living room holding his legs close together as if he were protecting his genitals. Let's not forget, a penis is not a phallus. Albeit the distinction (if not the object) is a slippery one, but for the purposes of abstraction and not to prematurely divest the phallus of its symbolic importance, the phallus remains disarticulated from the realm of the physical body. Moreover, there is no point in fetishizing the concrete in this instance. Rosie enters the room carrying a gun; or rather, Rosie enters the room brandishing the phallus. Like any good revenge fantasy, the instrument of destruction is turned on the destroyer himself, only in this case that instrument of destruction has undergone a transmutation into a symbolic equivalent. Rosie points the phallus/gun at David, while Lily, who has remained complicit in her loyalty to the heterosexist institutions of phallocracy, reaches for a vase – no less than another symbolic equivalent – and putting the womb as a site of production into practice, she hits Rosie over the head with it – the vase that is. The vase/womb shatters, the phallus/gun drops to the ground. Before the referent has time to slip from her grasp, Lily recovers the phallus/gun and uses her newly found access to power to order a bewildered Rosie out of the house. Once Rosie has left, Lily relaxes, the phallus/gun slips away from her hands.

The lack of dialogue in this scene would appear to conform to a necessary silence of women within the logic of phallocentrism. In a somewhat uncanny replay, in theoretical terms, of the scene I have just recounted, Hélène Cixous elaborates this silencing in the following passage:

> It's a question of submitting feminine disorder, its laughter, its inability to take the drumbeats seriously, to the threat of decapitation. If man operates under the threat of castration, if masculinity is culturally ordered by the castration complex, it might be said that the backlash, the return, on women of this castration anxiety is its displacement as decapitation, execution, of woman, as loss of her head.
>
> (Cixous 1981: 43)

Lest we think of Lily and Rosie metaphorically losing their heads in their struggle, and Rosie as the object of a literal decapitation, which would leave us caught within the web of phallocentrism, it must be remembered that the vessel has broken into multiple fragments. And from those disorderly ordered shards it is as if an emancipatory potential had been released. The spectator might assume at the conclusion of this scene that heterosexual bonding remains intact, the woman-with-the-phallus, the woman who has put the womb as a place of production into action in order to temporarily acquire the phallus, loyally restores the vessel of phallocracy. However, in the following and penultimate scene of the film a significant reversal takes place.

We see a long shot of a police car driving up to the cabin in the early morning light. When two policemen – those exemplary men of the symbolic order – knock on the door, Rosie immediately assumes Lily has sent them to harass her and her daughter. Instead, the policemen inform Rosie that Lily cannot testify against her husband in a court of law, but what Lily wants to know is if Rosie would be willing to testify in her place so that charges may be laid. Rosie agrees. In the final scene, Lily shows up at Rosie's cabin with her children. She tells Rosie she has nowhere else to go. In a redemptive gesture Rosie accepts Lily into her home and the two women establish an alliance in their mutual, however discontinuous, fight against British imperial patriarchy. But this redemptive gesture on the part of Rosie, who must save the white woman, testify in court, and feel sorry for her, does not subvert the totalizing tendencies of a liberal feminist paradigm that would construct women's alliances on the presupposition of an essentially gendered solidarity.[5] We do not get outside "salvation history," to borrow a phrase from Donna Haraway, toward a hybridized figure of de-totalization.[6]

In order to create a friendship earlier in the film, Rosie and Lily had to overcome their cultural differences, characterized by racially oriented and socially inscribed class discontinuities. The parallel editing technique juxtaposes representations of Native and English domestic spaces and attributes to those spaces particular familial states of being: indifference, sexual abuse and emotional neglect within the English colonial sphere; love, friendship and maternal nurturing within the "other." Unlike their friendship, Lily and Rosie's alliance comes about through a violent set of familial circumstances which cause the women to unite in a mutual struggle against a common enemy: a Man who represents the phallic ordering of British imperialism. It is their gendering as "women" which creates the mutual basis of their alliance and hence the possibility of overcoming an essentially masculinized representation of imperial violence.

BEYOND THE HORIZON

While race and class differences are seemingly overcome through mutual understanding, the antagonistic, or what I would call discontinuous aspect of Lily and Rosie's relationship is centered, for Wheeler, on a unified notion of sexual differentiation. The centering of sexual difference as the site of colonial violence needs further elaboration if sexual difference is to be understood as historically contingent upon the varying social relations particular modes of production make possible. The mutuality that gender affords for their alliance masks the discontinuities of gender formation in the competing spheres of gatherer/hunter and capitalist modes of production.

105

While we have already noted the usefulness of Althusser's theory of ideology for an analysis of Wheeler's gendered narrative of British imperialist violence, I would suggest that what figures as a "relatively autonomous" ideological effect in this film can be described, more precisely, as a residual (or/and emergent) mark of competing material forces between or among differing modes of production.[7]

For example, under a capitalist mode of production sexual difference furthers the exchange of women's bodies whereas for gatherer/hunter societies, although sexual difference may determine a division of labor, it does not necessarily determine gendered divisions of power: who makes decisions and who is responsible, in general, for decision-making practices. Indigenous women are viewed as objects to be exchanged for the purposes of bearing political and tribal alliances, pawns in the peace-making practices of their male chieftains, fathers and brothers.[8] Little thought is given to Native women as political agents who make decisions about the efficacy of certain kinship affiliations, accepting some and rejecting others, depending on the needs of the larger community, except in terms of a patriarchal equivalence named matriarchy. It is only on the basis of the gender essentialism of Western knowledge that a biologically based phenomenon such as the sexual division of labor could serve as the sole indicator as to how power is distributed and deployed throughout a society. And further, it is only in the reified consciousness of late capitalism that women appear to be objects that can be exchanged with the regularity of beads or coins. In a gatherer/hunter mode of production, based on a reciprocal exchange of "objects," there is no essential correspondence between a division of labor based on biological sex – women gather/men hunt – and the terms of agreement as to how decisions are made and how power is disseminated, deployed and resisted (Emberley 1990-1; 1993).

In *Loyalties* the question of the relatively autonomous relationship between cultural differences and the political economy of competing modes of a dominant capitalist production and residual (if not emergent) gatherer/hunter production is represented in the figure of the Native grandmother. It is the Native grandmother who speaks the violence of the historical confrontation between British mercantile imperialism and North American gatherer/hunter social formations. She represents the trace of kinship and matrilocality. It is her home that becomes the spatial equivalent to Lily and David's newly built house. During a barbecue scene on the waterfront on the Sutton's property, the grandmother informs David, "I was born right across there in a house my grandpa built. We lived here for a hundred and fifty years, until the white people wanted to live by the lake. They told us we were slaughterers." The grandmother is the only figure to voice the historical violence of colonialism, and, as if to demonstrate the continuing descent of that violence, it is the granddaughter who is made to bear that colonial violence in its contemporary mode of domination.

I would like to conclude my discussion of *Loyalties* by returning to the Levall–Bedard case and the passing of Bill C-31 and its consequences for feminist theory and practices. Anne Wheeler's film is a multi-layered representation of the gendering of the history of imperialism as an originary site of patriarchal violence. The ideological effects of this representation of an originary violence against Native and non-Native women can be said to lend support to, if it is not already supported by, Anglo-Canadian women's involvement in the Levall–Bedard case and the eventual implementation of Bill C-31. In contrast, many male-dominated Native organizations opposed the case. The Assembly of First Nations opposed it on the grounds that resolving Native disputes within the Indian Act maintained the Act as a tool of colonial state intervention, thereby retaining the Canadian state's control to define who is and is not a "status Indian." In contrast, Justice Bora Laskin, one of the Supreme Court Justices who heard the Levall–Bedard case, argued against the majority decision, stating that the Bill of Rights (1960) should take precedence over the Indian Act.

While the eventual passing of Bill C-31 in 1985 provided an important precedent for testing sex equality under the new Charter of Rights and Freedoms, it also, not surprisingly, unleashed an extraordinary set of contradictions and clashes among Native organizations and communities. The involvement of various professional and white women's organizations in the passing of Bill C-31 has been criticized by many Native women activists, cultural critics, artists and writers, who have become increasingly sceptical as to whether liberal and socialist feminists understand the meaning of *difference* between the sex and gender formations within gatherer/hunter cultures and those existing and imagined within a late-capitalist ideal of liberal plurality and diversity. Since 1975, Native women writers, such as Jeannette Armstrong (1989), Maria Campbell (1983) and Lee Maracle (1988; 1989), have insisted that their cultures are neither the exclusive victims of a European fantasy of extinction, nor the redemptive heroines of a Canadian feminist fantasy of women's liberation from inequality.

As a response to the continuing contradictions of colonialism, *Loyalties* offers a utopian resolution: an Anglo-Canadian woman uniting with a Native woman against European patriarchy. The particular configuration of "sisterhood" that *Loyalties* represents is like an imaginary resolution to the gender-specific contradictions of colonialism. In attempting to address the question of the place of feminism in relation to Native women's specific struggles, Donna Haraway provides an important place to begin when she writes of the "daily responsibility of real [i.e. actual] women to *build* unities, rather than to naturalize them" (Haraway

1990: 153). That the postcolonial condition of Lily's freedom is gained at the expense of Leona's victimized body should indicate that the racial violence of colonialism is being mediated, and hence naturalized, by the language of sexual difference. This language is constructed in the film by Wheeler's parallel editing technique, used to represent a gendered class opposition. By situating the film in the social context of the political alliance-building which took place between the women's movement and Native women's organizations in the Levall–Bedard case and over the passing of Bill C-31, the specificity of the differences in gender formations between gatherer/hunter and capitalist societies is clarified, thus providing a basis to understand the mutual irreducibility of feminist and Native struggles, aims and interests. While solidarity between feminism and Native struggles for self-determination articulates the political horizon of liberal feminist expectations, *Loyalties* aesthetically represents the contradictions of this articulation in its depiction of a British woman's realization that an alliance with a Native woman is a necessary condition of her (that is Lily's) survival and freedom.

One of the liberative and appealing attributes of postmodern thought is its concern with "otherness," and the representation of colonized people and women, among others (see Introduction/Conclusion). The problematic of otherness not only directed attention toward the European or Euro-Canadian subject's need and desire to consolidate its subjectivity through the making of others, but it also opened a window onto the desire of previously disenfranchised groups to speak for themselves, in their own voice. *Loyalties* dramatizes this postmodern condition in its staging of an alliance between a Native woman and a transplanted European woman. The Native/Woman/Body constitutes a fractured identity formation and points to the divided loyalties the film thematizes and hierarchizes among a set of presupposed choices offered up by a post-sixties phenomena of "new" social movements: feminist, gay and lesbian, anti-racist and anti-imperialist struggles, for example. In Wheeler's liberal feminist narrative the resolution of antagonistic differences would appear to lie elsewhere; somewhere in decolonial space, a space that is neither the inner text, the productive womb of colonial violence, nor the outer text, the productive masculinized history of imperial expansion, but the space of decolonial movements and struggles that cannot be made to consolidate the primacy of a feminist project for global solidarity, however much Wheeler may wish this to be the case. In its political expression these choices are represented by an antagonism between feminist and/or decolonial struggles that must be resolved. The site of struggle to articulate the significance of these movements becomes, however, Leona's body; a Native woman's body that is sexualized and naturalized as a legitimate site for the resolution of social, political and economic conflict.

NOTES

1 Thanks are due to Peter Kulchyski, Laura Murray and Jane Parpart for their generous commentary and helpful suggestions in the process of writing this essay.

2 The history of Bill C-31 is told by Native women themselves in *Enough is Enough: Aboriginal Women Speak Out* (1987), as told to Janet Silman. This book clarifies the material conditions of Native women's oppression, as well as documenting the grass-roots political action of the Tobique women in their political and legal struggle with the federal government. For an account of sexual discrimination in the Indian Act see Kathleen Jamieson's "Sex Discrimination and the Indian Act." The term patriarchy was originally borrowed from anthropology (Bannerji 1991). The term was used to characterize indigenous cultures as uncivilized and barbaric.

3 Paula Gunn Allen, in *The Sacred Hoop: Recovering the Feminine in American Indian Traditions*, critiques the imposition of a European civil patriarchal ideology. She coins the word *gynocide* to describe the extent to which European patriarchal formations attempted to severely cripple, if not destroy, the position of indigenous women within gatherer/hunter social formations; for their position represented not only a threat to imperial state control in North America but also threatened the state's continuing subordination of European bourgeois women coming to the "new world" in the mid to late nineteenth century (see Sylvia Van Kirk 1980).

4 There is an important scene which establishes Lily's problematic relationship with her mother. Lily and Rosie are in a field gathering berries. Lily tells Rosie about her distant relationship to her mother, who died leaving her "pots of money" but few emotional ties.

5 Nowhere is the redemptive dream of feminist alliance-building with Native women's struggle more dramatically played out than in Linda Griffiths's (1989) account of the making of the play *Jessica*, which is based on Maria Campbell's autobiography *Halfbreed*. Griffiths recounts her struggle to work through her racism and guilt as she created the play. Campbell makes it clear that she has no interest in colluding with Griffiths's desire to turn her into a savior.

6 For Donna Haraway, the "cyborg," an animal and human organism/ machine hybrid, becomes an exemplary postmodern figure of de-totalization. As Haraway writes, in a paradoxical twist on Christian theology: "The cyborg incarnation is outside salvation history" (Haraway 1990: 150).

7 Althusser gives cultural and aesthetic objects a degree of "relative autonomy" from the economic base as the final determinant in capitalist structures of exploitation and oppression. Class relations no longer hold a privileged analytical place. While gender as an ideological determination has relative and not absolute autonomy from the relational exigencies of class positions, on the other hand, the processes by which gender is engendered constitute ideological effects that demand historically and culturally specific modes of analysis.

8 The most influential formulation of women as objects of exchange comes from Claude Lévi-Strauss's *Structural Anthropology* (1963).

6

COPING WITH MARGINALITY
South Asian women migrants in Britain
Suresht R. Bald

In the 1950s when Britain was trying to rebuild its war-devastated economy it needed labor. Long years of imperial history made it turn reflexively to its ex-colonies to fill its needs. Dislocated from their ancestral homes when the British partitioned India, thousands of Pakistanis and Indians responded by moving to Britain in search of a new life. By the late 1950s and early 1960s the South Asian migrants in Britain had reached a sizeable number, at least large enough to cause racial riots. The Conservative government, experiencing a downturn in the economy, claimed the migrants were "taking away jobs" from the indigenous people. Controls on immigration from the "colored" Commonwealth were proposed in Parliament. Fear of these controls led South Asian workers to bring friends and families to Britain *before* the 1962 Immigration Act went into effect (Anwar 1986: 9). These new arrivals flocked to the areas where members of their extended families and villages lived and worked (Anwar 1979; Jeffery 1976). The 1960s also saw an influx of Indian and Pakistani students to British universities, a number of whom stayed on as doctors, lawyers, teachers and writers.

The third wave of immigrants came in the late 1960s and early 1970s when people of South Asian origin were expelled by the newly independent states in East Africa. Currently approximately 3.5 percent of the British population is of South Asian origin. Forty-five percent, however, are British born. There is therefore no uniform South Asian migrant; significant differences exist among them.

These differences, however, are erased by the British who ignore the immigrants' diverse histories and only see what they consider similarities in skin color, speech and dress (in reality the South Asians' "color" varies as does their language and even dress). The British immigration laws, the derogatory term "Paki" (short for Pakistani) by which the South Asians are stigmatized regardless of their national origins, and the stereotyping present in popular fiction and media, all elide the differences that are important to the definition of the South Asian immigrants' identities.[1]

The difference that receives attention is that between the South Asians and the white British.

The imperialist ideology, which at one time was necessary to justify British colonialism in India, now represents the South Asians' "deviation" from the British-defined norm of physiognomy and cultural practices as evidence of their inferiority to the British. Graffiti on private and public walls demand "Pakis go home"[2]; newspapers, almost daily, include reports of "racial incidents" (euphemisms for fire-bombing of South Asian homes, assaults on South Asian children by their peers at school, "snatching" of gold chains off South Asian women's necks) and accounts of racial slurs/"jokes" made by someone in a position of power (see chapters 4 and 6).[3]

Though the countries of the subcontinent from which the migrants emigrated have been "independent" since 1947, two hundred years of colonial history continue to structure relationships between the British and the South Asian migrants.[4] As the colonizer, not only did Britain control the economy, politics and education of the people of India, Pakistan and Bangladesh, it also erased and rewrote their past to shape the present and future that met the colonizer's needs. The pathological nature of such a relationship is illustrated powerfully by Fanon in *The Wretched of the Earth*, who described colonial relations as[5]:

A world divided into compartments, a motionless, Manicheistic world . . . ; a world which is sure of itself, which crushes with its stones the backs flayed by whips: this is the colonial world. The native is a being hemmed in; . . . The first thing the native learns is to stay in his place, and not to go beyond certain limits.

(1963: 51–2)

These "limits," which are both mental and physical, are often internalized by the colonized, who then "instinctively" know what they can or cannot do, where they can or cannot go or live. Thus their marginalization maintains the center.

In the case of South Asian women migrants, the delineation of the limits is done both by the dominant group, the white British, and the patriarchal religious and cultural practices of the migrants' homelands. For while the white British construct the boundaries between the center and the margins and decide who resides where, it is the expectations of the women's own religious and cultural reference group that define *how* they live within these constructed margins.

The postcolonial condition has received much attention in recent years by scholars who themselves are postcolonial.[6] However, voices of migrant women who traditionally had been silent/silenced in the literature on Western politics and society are still rather soft in the postcolonial discourse.[7] This study is an attempt to increase the "volume." Based on extensive interviews with South Asian women from different class, age, national origin and religious groups, the study provides a first-hand account by the women themselves of their lives as migrants in Britain, the land of their ex-colonizers. By focusing on their experiences within their families/homes, their workplace, their particular South Asian community and the larger British society I hope to reveal the remarkable courage, inventiveness and zest for life with which these women have created new communities to replace the ones they left in their native lands, the strategies they have fashioned to empower themselves so that life in the "margins" is more bearable, and how the interaction of the cultural and religious practices of their homelands and of British reality have constructed and constricted or empowered their selves. I argue that while British racism and institutions, plus the patriarchal nature of Pakistani, Indian and Bangladeshi family structures tend to impose an artificial sameness on the diverse histories of the South Asian women immigrants, the lives of these women differ depending on their particular location in the social and economic hierarchy of the land of their origin, age, their place within their families (daughter, wife, widowed mother), their work and the particular cultural and religious practices which prescribe their lives.[8] Even their status as "migrants" is problematic: South Asian women who came to Britain after their husbands or fathers were expelled from Kenya and Uganda in the late 1960s and early 1970s were categorized as "migrants," despite their British "nationality" just because their skins were of the "wrong" color; and the British-born South Asians, likewise, are considered "migrants" even though they know no other home than Britain.

However, despite the importance of the differences among my informants, it makes political sense to maintain the category "South Asian women migrants in Britain" when discussing the women as a "group."[9] For though their different subject positions may define particular responses to their marginality, their categorization as "Pakis" by the dominant group has led to coalition-building on specific issues.[10] As "Pakis" they come together as Asian women.

CLASS AND MARGINALITY

The world of Leena Dhingra, a published writer of fiction, who speaks flawless English with an educated accent and lives in the posh north London suburb of Hampstead is very different from that of B. Kaur who works in the restrooms at Heathrow airport and lives in Southall. Their locations in the Indian and British class structures affect the understanding of their marginality and the strategies they choose to deal with it.

Leena Dhingra is in London because her parents lost their family home in Lahore in 1947 when Pakistan was created out of then British India. A decade or so later the Dhingras acquired a home in Hampstead in order to be near Leena's father who was working for UNESCO in Paris, and yet where people spoke English and the children could obtain schooling. "I came to England when I was eleven . . ." Leena tells me. But already she had lived in Switzerland, India and France; and in each country she attended private schools. At the age of fifteen she dropped out of school.

> I was quite unprepared for the real world. In the English private school I had never encountered racism; but as a fifteen-year-old trying to find a job in London I met it everywhere. I think there is always an element of racism [present in interaction with the white British] . . . I am not talking about overt racism . . . it's this kind of thing about not being quite right, not being kosher. There are lots of things in which you are different, you are just different, but there is a way in which this difference is not quite right to be different, yet there is nothing you can do about it. So people [the Asians] try to disappear, to be less different as they possibly can. It's a funny kind of thing. It's a kind of non-acceptance. . . . It was very difficult for me. There were lots of things that I could not digest. In my schooling years – I'd been to a prep school – it [racism] wasn't anything I had *ever* encountered. . . . So when I came here [London] it was a different kind of a world. I could not understand why there was this rejection. I had been to India and I was aware of where my family was situated in terms of class. We had status and privileges; and then to be in this [inferior] kind of a situation here . . . it was very difficult . . .

In London neither Leena's family name and status, nor her public school education protect her from covert racism. She is grouped with other South Asian immigrants; she is a "Paki," no different from the South Asians who provide services that the white English no longer consider worthwhile performing themselves. The sense of loss that she feels so acutely reflects the loss of privileges with which she grew up; she feels demoted socially.

B. Kaur works in the restrooms at Heathrow's Terminal Two. She is a middle-aged Sikh woman who looks older than she is. She wears the Punjabi baggy pants and long shirt dress (*salwar kameez*). She sits on a stool in the restroom and after a toilet has been used she goes into the stall to check the toilet paper supply and general cleanliness. Occasionally she wipes the toilet seats with a rag dipped in an antiseptic solution. It is work that would be done in the Punjab by the low caste sweeper whose "touch" was traditionally considered "unclean." B. Kaur is from a caste much higher than the job she is doing, but she is not disturbed by this "demotion." Her meager earnings go into the family "kitty" (pool), which her husband manages.

Conversing with me relieves the monotony of B. Kaur's job. She talks freely in her native language, Punjabi.

> "My husband came here to work in the brick factory. He has worked in different factories. Now he works in the bakery (at Heathrow). After being here two years he called me . . . almost 25 years ago," she tells me. "I have two sons and one grandchild. I come to Heathrow at six in the morning and leave at two in the afternoon. We get a half-hour lunch break. The work is alright. There are quite a few of us. . . . Most of the other women who work here are from the Punjab though there are a few from Bangladesh and Pakistan. We get along well. When it is slow we visit each other's restroom stations. Most of us live in Old Southall though some live in Ealing or Southall. We own our house . . . my husband has fixed it up nicely."

As we are talking a couple of women come and want to know where they can leave their bags. B. Kaur responds, "Yes, yes, go." Perplexed by this response they ask her if she understands Spanish. I interject, translate the tourists' questions and B. Kaur's answer which directs them to the left luggage. This interchange prompts me to ask what language Ms. Kaur's children speak at home. I get a quick and curt reply:

> They speak Punjabi, that is our language. I tell them that I do not care what they speak outside the house, but once they cross my threshold they can only speak Punjabi. I have good boys, they listen to me. Life is not easy here, as you know. We have to work very hard, but I'm not afraid of hard work. What I don't like is the cold. The winters were cold in our village but it was different, it was not this wet cold. I don't think I'll ever get used to it.

Unlike Leena, racism is not uppermost in B. Kaur's mind. It is the boredom of her work, the preservation of her language and cultural practices, the need for economic stability, and London's harsh winters that constitute her story. The reasons for her emigration were economic and in her judgment she and her husband have been successful. In contrast to Leena, B. Kaur feels she has improved her status because she now owns her home and her children have white-collar jobs. Born in an Indian village where life was hard, she is accustomed to living in the margins. Like the colonized natives in the Manicheistic world described by Fanon, she has learnt "not to go beyond certain limits." Like other women from Punjabi villages I met, who came to England to join their husbands or sons in the late 1950s or early 1960s, B. Kaur has accepted the divisions between the center and the margins.

An example is Mrs. Singh who lives in Southall and whose husband also works at the bakery. Mrs. Singh does not speak the language of her adopted land though she has lived in England for almost thirty years. Indeed, by not learning the English language these women and the men in their lives have ensured that the sharpness of the differences between the British and them is maintained. Of course it also keeps the women dependent on the men.[11] These distinctions are strengthened and reproduced by living and working as she does among her own people.

Southall, a working-class "suburb" in west London, which was one of the few areas where the South Asian immigrants could live in the 1950s and 1960s, has now been territorialized and transformed by them into a little India/Pakistan. The dress shops display saris and *salwar kameez*; jewelry shops carry 22-karat gold jewelry catering to the South Asian taste; some butchers sell *halal* meat (meat from animals slaughtered according to Koranic injunctions) while others sell *jhatka* (meat from animals slaughtered according to the rules of Sikhism). Indian fruits and vegetables, spices, savories and sweets, including beetlenut are available at reasonable prices. Most of the people in the stores and on the main streets are South Asians. I notice that here, among themselves, the South Asians, male and female, walk and talk with assurance and dignity. Southall has been transformed from the "margin" into a "center." Outside Southall and other similar areas, they walk drawn within themselves to occupy the minimum of space.

Unlike the South Asian residents of Southall, Indians like Leena Dhingra have to be on their guard all the time. Despite their educated posh accents and middle class, Anglicized life-style, it is not easy for these women to enter the "center." Indeed, the "center" not only does not accept the Anglicized Leena, but continues to stereotype and "label" her as a "Paki," erasing the difference between her and B. Kaur, who cannot speak English and holds an unskilled job. In her writing Leena responds both to her "non-acceptance" and her stereotyping: "I have taken these experiences of racism and created a short essay called 'Labels.' I don't directly deal with racism [in the essay] because that is not so effective, but the essay is about how we are labeled and put in little boxes that confine us."

BEING ELDERLY, "ALIEN" AND WOMEN

For the non-working South Asian elderly women who emigrated late in life, migration means leaving the protective and familiar environment of the village for a society whose culture and language they may not understand. This transition often includes leaving a large extended family for a small nuclear one, and disrupting marital relations developed within the extended family context. In alien surroundings these women have created new communities to replace the ones in their motherlands; they have learnt new ways to interact with their husbands, and creative ways of combating loneliness.

Saleema lives with her husband in north London. She came to Britain after Bangladesh became independent and when non-Bengali Muslims felt it safer to leave. She is now in her late sixties.

> When I feel lonely, [she says] I catch a bus. . . . I go to the end of the line and return. . . . For three hours I'm in the midst of people and activity. I don't talk to them, but they are there. I miss the comings and goings, the intense activity that always went on in our family in India and Pakistan. I love people; I miss not being in the midst of a lot of people.

The London bus, a substitute for a Muslim joint family household, keeps her sane. I ask "Do you have friends here? How do you spend your time?"

> I belong to one group that has twelve women; we meet once a month in someone's house. Then there is a group in Tooting [council-run center for the elderly] that meets once a week. But I haven't been there for six months now. Hindus, Punjabis, Gujeratis, my community women, and those who came from Africa, all South Asian women attend these meetings. They [the council] give us lunch and there is generally a speaker on health or nutrition or something; . . . I used to go there every Wednesday, though I haven't since I found another place close by. I go to this new place three times a week.

The artificial community of these council-run programs helps fill Saleema's need to be with other women. In some ways the artificiality of this functional community is preferable to her, for unlike the community of which she was a part in Pakistan the women she meets at the centres for the Asian elderly are neither judgmental nor demanding.

> If there is nothing going on I just go to Bingo, [she continues]. I enjoy Bingo a lot. I get peace of mind there; whenever I am upset I go there. . . .
>
> After breakfast I read the *namaz* [the Muslim prayer] and the Koran, then I exercise. After that I cook the meals, sometimes I cook for the whole week, and leave the house. I return in the evening. I do not like being in a quiet house. . . . When I was young in Kathiavar our house was next to our cousins and aunts and uncles. . . . There was a great deal of activity. . . . I feel strange

116

by myself. My husband's nature is very different . . . he has never given me any companionship. I can't be by myself; if I am alone I turn on the television. I need someone around.

Back at home in her Muslim joint family Saleema saw her husband essentially at night, in bed. In London, where she and her husband live alone in a flat, she has worked out creative ways of minimizing her contact hours with him without abandoning her responsibilities for cooking and cleaning.

By contrast, Mrs. Singh (mentioned earlier) is a Sikh. She lives in a two-story row house in Southall on a clean street of well kept houses. Her daughter lives nearby and there is a Sikh *Gurudwara* (place of worship and social gatherings) within easy walking distance. Unlike Saleema, she is surrounded by family and friends. When asked about her day she responds, "I take care of my grandchildren. I am active in the *Gurudwara* and we older women meet for *satsang* [meetings where women sing religious songs] once a week. But to tell you the truth there is enough work in the house and I have always someone dropping in."

In the Sikh community the *Gurudwara* provides a haven for the elderly who are either no longer or never were in the workforce. It is a substitute for the Indian village community, where various festivals and ceremonies are held. The women cook a hot meal at lunch which is then served free to whoever wishes to eat there. These activities provide occasions for socializing with other women and relieve loneliness and boredom. For Hindus, the Sikh *Gurudwara* have their counterparts in Hindu temples. But in the lives of older Muslim women like Saleema, the mosque does not play a similar role. Saleema, therefore, has to create or search for other alternatives.

Hindu and Sikh tradition demands that the sons take care of their aging parents. But the reality of life in England has made this difficult: "Our sons see us as a burden. It is so expensive here; and houses are small. As the children grow older they want their own rooms. We are seen as taking their space. . . . We are totally dependent on our children," explains one elderly woman at the Streatham Centre for South Asian Seniors. "We have nowhere to go." She is corrected by another, "We can go to the shelter [Asra, shelter for South Asian women] or ask for a council flat. If we want to we can be on our own but we are afraid that if we do so our family's name will be mud. So a lot of us endure abuse without protesting. We try to keep ourselves busy and out of our children's way."

While the sons feel free to flout tradition, because of long years of socialization the widowed mothers do not want to sully their son's name in the community. For the honor of the family is tied to its members' proper fulfilment of their duties and responsibilities. And it is this "honor" that plays an important role when marriages are arranged. For the sake of their grandchildren, therefore, the mothers must keep up appearances.

BEING YOUNG, "ALIEN" AND WOMEN

Loneliness is not the bane of the elderly alone. Young Anjana, who came to Britain from Kenya at the age of seventeen, remembers her years at the university:

> There were very few Asians at the university in Leicester; I remember only about ten or twelve faces that were Asian. I am talking of 1974–7. I felt very isolated. It was difficult for me to mix with the white students. Most of them lived in the hostel; they tended to go out with each other and party but I couldn't because I was living at home, and I had to deal with family restrictions: "you must be in you can't go out."
>
> Did that bother you?
>
> It did sometimes but what can one do? It's a fact of life. That's how our families are.

Ravinder Randhawa, who is a writer of fiction like Leena, tells me,

> I went to a grammar school where I was the only Asian girl for a while. I was lonely at school until another Asian girl came. Then I realized how wonderful it was to have a friend to whom you did not have to explain. Who accepted you as you were. After grammar school I went to a polytechnic. There I met other Asian students. It was very interesting how we all came together. . . . I felt much freer with my Asian friends. I didn't have to always explain . . . we had shared experiences.

Ravinder's father has worked at various factories, and her mother, when she worked, was a house-cleaner. Having struggled all their lives for the economic wellbeing of the family they place a higher premium on economic success than their university-educated daughter:

> I've always felt very close to my parents. But inevitably there were certain things that were difficult for them to understand. I was studying English. I always knew I wanted to write. . . . When I told my parents they thought I was crazy. They wanted me to have a proper profession. So I agreed to do a teaching course. I taught for a while . . . but I really hated it so I quit. Then I got a job organizing Asian women workers. I enjoyed that work. . . . But I had to prove to my parents that I could make it on my own. So I worked and

saved money to buy a house. I rented out rooms to friends so I could pay the mortgage.

Having proved her ability to survive Ravinder won over her parents to her choice of writing as a career. She did not defy her parents, rather she worked patiently to gain their assent. Her success suggests ways in which the younger generation "immigrants" have effected change without becoming alienated from their families.

MARRIAGE IN AN ALIEN LAND

For the young South Asian women who were either born or grew up in Britain, marriage/family are where struggles for empowerment are enacted. While Ravinder (above) was able to influence her parents, another young woman, Harbinder, gave in to her parents' choice for her husband. Harbinder was born in England and speaks English with a thick south London accent but her Punjabi is flawless.

I've never visited India. I work in a building society. My husband works in the bakery (at Heathrow). Most of the workers in the bakery are Punjabis. . . . It was an arranged marriage. My parents showed me his snap (photo) then he came to London and I met him. He stayed with us for a couple of months and I agreed to marry him.

What if you had said no . . . could you have said no to the marriage proposal?

I guess if I really didn't like him my parents would have listened to me . . . but it is difficult, you know, he is a relative, a far off cousin, so we knew the family, how could I say no? . . . He is a good man. He does not beat me or anything; he is good to me. But he has problems. You see his English is not very good. . . . He works with other Punjabis and they speak only Punjabi at work. I think he has even forgotten the little bit of English he knew when he came here [nine years ago]! Which means that I have to deal with the outside world for him . . . sometimes I really resent it. Like today. Its very time consuming, and I work and also take care of the house and cook etc.

Does your husband help with the housework?

He does a bit. But both of us work. . . . But he is quite good, he helps with the dishes.

Harbinder's fluency in English, and her husband's unintelligible accent have reversed the traditional husband–wife relationship: instead of the husband negotiating for the wife in the public sphere it is Harbinder's task to represent her husband. The role reversals are evident also in the nature of their respective jobs: Harbinder works in accounts in a credit union (typically men's work) while her husband works in the kitchen – the bakery. The fact that she works, and in fact earns more than her husband, renders problematic the role of men as primary earners in their household. Her arranged marriage has lasted, but she admits sometimes feeling resentful of all the pressures placed on her.

Surinder is an example of a marriage in which the bride was imported from India. She grew up in a village in north India; the marriage that brought her to England ended in a divorce. She speaks Punjabi in her village accent.

> I came here as a bride. My husband was settled here and went to India to find himself a wife. The family thought it would be a good match, and were they wrong! The moment we came to London he became very abusive. First it was just his language but then he started beating me. I took it for a while but then decided enough was enough. . . . I went to a lawyer who was very helpful. He helped me get a council flat and later I got a divorce.
>
> Was the Indian community helpful?
>
> [She looks me straight in the eye] I didn't need friends, I needed a lawyer. And I found one.

It took courage and determination for Surinder to get out of an abusive marriage. But it helped to be free from family pressures she would have been subjected to in her village where marriages are meant to last not only in this birth but in the next seven![12] Since Surinder speaks mostly Punjabi I ask how she survived after she left her husband.

> I found work as a seamstress in a factory. I worked for several years till my back could not take it any more. I don't work now; I get disability benefits. She pauses, then repeats, I have a council flat. I'm self sufficient; I don't need my husband's abuse. . . . Here I have my disability pay, I have a flat, I have health care. . . . In India I will have to live with my family, I'll be a divorcee daughter or sister who has to be taken care of. I'd lose my independence and just be a burden to everyone. I'm happy here.
>
> Is this quite common, men going to India for a bride and then mistreating them?
>
> Unfortunately, it is common. . . . When it comes to marriage the men want a wife who has been brought up in India. They think that the women who have grown up in England are too bold; they want a wife who will listen to them and do as she is told.

Anjana the daughter of a Gujerati shopkeeper who migrated to Britain from Kenya, speaks fluent English, spiced occasionally with choice Hindi or Gujerati words. Her husband was settled in the United States and he went to England in search of a bride.

It was an arranged marriage though we did meet each other. He was a computer programmer for Lockheed living in Atlanta, Georgia. Though he was from our community, the family did not really check him out. They figured that he was well educated, he had a good job. . . . But he turned out to be miserly and mean. He had this strange notion that all women were after money. . . . He would give me $12 to spend on food for the week and for that too he wanted a strict accounting. I was quite miserable. I had no freedom to go anywhere or buy anything that he did not approve of. . . . So I left him and returned to England. The breakup of my marriage was very unsettling. I went through a period of depression.

Clearly arranged marriages are the norm among working- and lower class South Asian immigrants. However, incidence of divorce is much higher in Britain than in comparable communities in India and Pakistan. The cultural practices of the host country, the availability of social services (stipends, low rent council flats), and in Surinder's case her distance from her immediate family and community, make divorce easier. However, not all women avail themselves of divorce. Gurdip Kaur Sandhu, whose case became a cause célèbre for South Asian women, endured abuse at her husband's hands for seventeen years and was finally murdered by him and her brother-in-law on 11 May 1986.[13] This case served as a catalyst for bringing together women across class, religion and national origin for demonstrations against male violence.

Though the older generation Indians see the continuation of arranged marriages as another way of drawing the boundaries between themselves and the British, the margin and the center, some women are beginning to resist the practice. Ravinder met her husband while working as an organizer of Asian women workers. "At first my parents were not too happy because our families are quite different. My father-in-law is an engineer. But now they have accepted everything and they are very supportive." Leena chose her own husband. "I married a man whose family had lived in the same village in Belgium for generations. The stability of his family attracted me; I hoped some of it would rub off on me. But I ended up uprooting him," she laughs. Her marriage too has ended in a divorce.

WORK/INDEPENDENCE/EMPOWERMENT

Work outside the house is empowering for middle-class South Asian women who had never worked in their homelands. Two examples are Saleema (mentioned above) and Mrs. Ameen. Mrs. Ameen's husband had a flourishing retail business in Nairobi. Though the Ameens were able to bring some of their money to Britain when they emigrated, their life underwent a radical change.

There were no servants to help me. I had to do all the housework *and* help at the shop and sub-post office that we bought when we came here (London). We lived in the flat above the shop. I had never worked before, but I found that I really liked working in the shop. In fact I liked it better than housework! [She laughs.]

For a few years, 1974–7, Saleema, who came to Britain from Pakistan in 1972, ran a restaurant with her daughter.

I was happiest when we had the restaurant here and I did the cooking. It made me feel worthwhile; it made me forget my pain. The restaurant kept me so busy that I hardly had time to eat, leave aside time to think. At the end of the day I would be so tired that I would hit the bed and fall asleep. In the morning I'd wake up get ready and start working in the kitchen. It was wonderful.

At the Streatham Centre for South Asian Seniors the group of elderly Gujerati women want me to listen to the story of Mrs. Patel, who was a school teacher in a small town in eastern India. After Mrs. Patel was widowed her son and daughter-in-law invited her to live with them. "My grandchildren were little at that time so I was needed. When they were old enough to take care of themselves my son's behavior changed. He became abusive. . . . One day he was really cruel so I packed up my bags and left." She went to a shelter for the Asian women called Asra (Hindustani for support) and has been living there now for two years. She says "I'm very happy there; I work in the Asra office for which I get paid. For the first time in my life I am independent, and I like it. I think my son did me a favor."

Judging from the respect and support the women gathered there give Mrs. Patel it seems as if she has been able to do what perhaps a number of them would like to do.

But not all work is empowering. For people like B. Kaur, working in the restrooms at Heathrow is neither rewarding nor liberating. Whatever she earns becomes a part of the family income which is managed by the husband. Moreover, she is dependent on the males from her community to drive her to and from work. Whereas Mrs. Ameen and Saleema were self-employed when they worked, B. Kaur is an unskilled wage-earner working under male supervision.

FIGHTING FOR EQUAL TREATMENT: NEGOTIATING DIFFERENCE

Anjana, the coordinator at the Centre for South Asian Seniors, has lived in Kenya, India, the United States and Europe. She feels most at home in England though she considers British society racist. She represents the new breed of young South Asian Britons who are secure in their own identity as Asians, while they fight for their rights and remain British in their outlook. Indeed, it is their belief in the British sense of justice and fairness that prompts their demands for rights equal to those enjoyed by their white counterparts. Anjana's energies are directed towards making Britain a better place for women, the poor and the elderly regardless of their color or national origin. She speaks with passion about the plight of the South Asian Britons:

> We had to fight for this center. It is quite unsuitable. . . . We have been after the council to provide better facilities. . . . A typical center for seniors in a white neighborhood has a large activities room on the main floor, a kitchen, television, a pool table, darts and a minivan. We have one old television. They [the Council] cannot understand why our people need separate facilities. . . . What they don't realize is that these women cannot speak English and, more importantly, they would never be in a place with strange white men. Most of the activities at these centers are for mixed groups. . . . Recently there have been several attacks on Asian women. A woman in a sari is very visible; if she is elderly she is also easy prey. There have been cases of sari-clad Asian women being willfully pushed as they get off the bus; they have been attacked waiting for the tube.
> . . . [W]hen an average British white person steps out of her house she doesn't have to fight for every little right as I have to . . . when I go out in the white world I don't get the same rights as the white British do. If I take back [to the store], say, a blouse I bought yesterday the salesperson will ask me a lot of questions, she'd be suspicious, and mind you, I speak English. Imagine the plight of the poor elderly Asian lady who doesn't know the language. . . . Having said all this, I *am* British, I take part in elections, I get involved in local causes. For example I am fighting for pensioner's rights for *both* white and Asian elderly. I recognize that British society is racist. . . . But *I* don't have to be racist. I work for the underdog wherever and whoever he or she is. Actually it is generally a she. [She laughs.]

Anjana's determination to fight for equality, to gnaw away at the center, if you will, in order to erase it, has helped free her of the depression that overcame her after her divorce. She speaks and acts her mind.

CONCLUSION

> They [the white Britons] do not see us as people, as individuals; they see us as Asians, and of course they already have a good idea what Asians are like, right? We just have to fit the mould.
>
> Anjana, coordinator, Centre for South Asian Seniors

The women's stories related in this study explode the notion of a uniform "South Asian migrant woman"; instead we see how differences of age, class, level of fluency in English, and religious and cultural practices define the women's understanding of themselves as subjects living in the margins of an alien society. However, even though their different subject positions define the sites and forms of their struggles for self-empowerment, all the women do share the struggle against the racism and loneliness to which they are subjected *because* they are brown migrants living in contemporary Britain. Their experiences are similar to many other migrants (and indigenous peoples) in the North (see chapters 4 and 5).

Migration took these women from the protective, familiar environment of their villages and extended families to the margins of a society whose culture and language some of them did not understand. In their struggles to cope these women have tried to create spaces for themselves; they have become more independent and self-reliant. Mrs. Patel's work at the shelter, Asra, has given her a sense of self worth, as have Surinder's determination to free herself from her abusive husband, Mrs. Ameen's work in the family shop and sub-post office, and Saleema's successes at her restaurant, and her delight with the London transport system and bingo.

However, for some of them, this self-reliance and independence is limited. Like the colonized natives in Fanon's Manicheistic world, women like B. Kaur have learnt "not to go beyond certain limits." These limits, imposed by the dominant group, have been reinforced for them by the religious and cultural practices of the lands of their origin. Their lack of education and facility with the English language, which isolates them from the dominant society, has also led them to create new communities which provide space for their religious and cultural identities. They live and interact with their own people, participating in practices of arranged marriages, *satsang* (communal singing of religious songs), reading the *namaz* (Muslim prayer), and visiting the *Gurudwara*. In the new communities they maintain and reproduce the distinctions between themselves and the dominant group, but they also minimize their loneliness and sense of isolation. The emphasis on traditional practices, however, also tends to equate the family name and honor with adherence to the culturally prescribed code of behavior. When Surinder needed a divorce, she knew instinctively that divorce may not be perceived as appropriate behavior by the South Asian community; she went to a lawyer, not her South Asian friends. Harbinder agreed to her arranged marriage because otherwise it would have been awkward for her family.

Migration has created novel situations within some traditional families. Harbinder's knowledge of English has made her Punjabi-speaking husband dependent on her. By contrast, B. Kaur's dependence on her husband has increased because she does not speak English.

Class differences separate the women who work in marginal unskilled jobs and the wealthier South Asians. "Indian" Leena and "Pakistani" Saleema who, unlike B. Kaur, Harbinder, Surinder and Mrs. Singh, live in the almost all-white wealthy suburbs of north London feel isolated and lonely. Differences in age, knowledge of English and degree of Anglicization, however, give a different meaning to each one's loneliness. While the well-educated, Anglicized Leena's problems are class related (she feels, given her background she has a right to be included in the center), Saleema's are rooted in the changed circumstances of her family. It is the loneliness of the nuclear family that drives her to creative alternatives for filling her need to be with people.

Differences of location, however, are erased by the covert racism all South Asian women experience. Some protect themselves from it by minimizing contact with the British, others either suffer or resist. Leena's and Ravinder's resistance takes the form of fiction in which the human costs of racism are presented. Anjana battles with the powers that be for equal treatment of the South Asian seniors, regardless of their national origin, religion, or class. By fighting against discrimination these women are trying to force the indigenous British to reject Fanon's Manicheistic world and instead accept and celebrate the pluralist, multicultural, multiracial and multilingual nature of the evolving British nation. They are fighting for change which they hope will erase the margins and thereby the center. This is a similar battle to the one being fought by women in the South who reject Northern representations of the "Third World woman" and call for the recognition of women's diversity, their unique experiences and their multiple, complex lives as well as their often common struggles against patriarchal values and structures.

NOTES

1 In my article "Images of South Asian Migrants in Literature: Differing Perspectives" (1991; 416–19), I point out how Anglo-British fiction writers tend to cast South Asians as either invisible or in stereotypic roles of anonymous restaurant-owners and workers, shopkeepers, or people who live in run-down houses and of whom "respectable ladies" are fearful and distrusting. The one character who does have a name and a face, Ali in Dennis Potter's *Singing Detective*, is portrayed as simple and childlike, much like the "native" of the imperial era.

2 In 1987 I was surprised to find such a message along with the swastika symbol on the door of the London building occupied by the Islamic Union of Britain.

3 This observation is based on careful reading of two British papers, the *Independent* and the *Observer* while in Britain for nine months in 1986–7, three weeks in 1989, and the spring of 1990.

4 See Bald (1991) for the persistence of imperial attitudes towards the once colonized South Asians.
5 See Fanon, *The Wretched of the Earth*, especially pages 35–106.
6 See, in particular, Homi Bhabha (1983; 1990: 1–7, 291–322). See also Spivak (1988: 179–268; 1990: 103–17) and Mohanty *et al.* (1991).
7 Amrita Wilson's (1985) and Susheila Nasta (ed.) (1991) are noteworthy exceptions.
8 See Mohanty for an excellent brief against universalizing definitions and for recognizing the "massive presence of the difference that our recent planetary history has installed" (1991a; 1992: 87).
9 For an excellent discussion of essentialism and difference see Diana Fuss (1989).
10 A prime example of such coalition-building was the campaign led by women across class, region and national origin, to demand justice for the murder of Gurdip Kaur, a 34-year-old woman, by her husband and brother-in-law in May 1986. The demonstrations against the men's acquittals suggest a similarity to what Spivak calls "strategic use of positivist essentialism in a scrupulously visible political interest" (1988: 205). Bernice Reagon (1983: 356–68) argues in a somewhat similar fashion when she advocates coalition-building across race for achieving specific political ends.
11 An excellent "fictional" account of this can be read in Rukshana Smith's novel *Salt on the Snow* (1988).
12 In North India every year married women fast on the fourth day of the full moon in October so that they may marry the same man in their next seven lives (rebirths). Even women with abusive husbands fast.
13 See issues of the weekly *Asian Times* (London) for May 1986–March 1987.

Part IV

THE RELEVANCE OF POSTMODERN FEMINISM FOR GENDER AND DEVELOPMENT

As is clear from the previous sections, postmodern feminism has raised some important issues for the field of Gender and Development. Yet, there is still much debate about the potential contributions of this perspective. Postmodern feminism is often criticized for being just another Western invention and imposition, too involved with Western concerns to have much relevance for women in the South. The complex and often inaccessible jargon of postmodern feminist writing seems out of place when considering the problems of women who are barely able to eke out an existence, much less enter academic debates. The emphasis on difference and diversity rather than unity is seen as undermining efforts by (women) activists to organize a broad women's movement to improve the lives of women in the South. At the same time, the previous sections reveal a need to rethink many of the categories which constitute the foundations of development theory and practice. Postmodern feminist critiques of Enlightenment thinking, their focus on language and power and their emphasis on previously subjugated knowledges have much to offer this process.

As we do not want to bring a premature closure to these important debates and issues, we have asked the three contributors to this section – Eudine Barriteau, Maria Nzomo and Mridula Udayagiri – to critically assess the (possible) contributions of postmodern feminism for Gender and Development, using their own experiences as a starting-point. Their different approaches can be partially traced to the different experiences and places from which each author writes, acts and thinks. Eudine Barriteau writes, for instance, as a Caribbean feminist scholar and development practitioner. She discusses the need to "Caribbeanize" social science theories which are generally imported from the West. She is rather positive about the role postmodern feminism can play in this respect. In contrast, Maria Nzomo is more critical of postmodern

127

feminism. Writing primarily from her position as a Kenyan feminist activist who has a prominent position within her country's democracy movement, she feels that postmodern feminism undermines feminist ideals and makes concerted political action more problematic. Mridula Udayagiri agrees. Relying on her experiences with grassroots work in India and her current participation in US academe, she seriously questions postmodern feminism's potential for political activism.

Despite their different assessments of postmodern feminism the three authors agree on one thing. Postmodern feminism can only be relevant for the field of Gender and Development if it succeeds in showing its potential for political action. Barriteau is more optimistic about this than either Nzomo or Udayagiri. She discusses the possibilities of drawing on postmodern feminist analysis to construct a Caribbean feminist theory that does not rely on androcentric or Eurocentric assumptions and concepts. She believes postmodern feminism can inform the construction of a Caribbean feminist theory that allows for multiple identities, diversity, multiple realities and does not set out to "victimize" Caribbean women. Postmodern feminism puts both political action and agency into perspective and reveals multiple political activities by women which might otherwise have gone unnoticed. In sum, Barriteau argues that Caribbean feminist theory, informed by postmodern feminist thinking, provides a way to connect seemingly disparate events and issues. However, Barriteau fails to address one important question. Is it possible to formulate a non-androcentric and non-Eurocentric Caribbean feminist theory using as an important building-stone postmodernism, which was first formulated in Europe and which has been under attack by feminists for its androcentricity?

Maria Nzomo and Mridula Udayagiri are less sanguine about the potential of postmodern feminism for Gender and Development issues. Nzomo's critique of postmodern feminism centers on its inaccessibility in terms of the language employed. This is also an issue for Mridula Udayagiri, who reminds us that women's literacy rates are still very low. To use highly academic language under those circumstances (re-)creates hierarchies among academic women, who have been mostly educated in the West, and women with only a few basic literacy skills, who might be at the forefront of local women's struggles.

Nzomo also criticizes postmodern feminism's overemphasis on discourse and representation at the expense of women's political and economic concerns. She thinks that representatives of concrete political struggles, like the democratization movement in Kenya, need to be able to appeal to universals especially in the face of repressive government practices. These universals lend additional legitimacy to a particular political struggle and can also be the foundation for the strengthening of ties with similar groups abroad. A good example of this is the alliance created between the Brazilian rubber-tappers movement and the international environmental movement; together they mounted enough political pressure to have the Brazilian government reconsider its Amazon policy.

Udayagiri largely agrees with Nzomo. She doubts the political potential of postmodern feminism, fearing that its excessive emphasis on difference and contextuality undermines the possibility for women's movements to effectively struggle for change. Moreover, she challenges postmodern feminism for its focus on essentialism/universalism, which in her view precludes other political implications from its analyses and for its silence about the political nature of counterdiscourses or practices of resistance. In other words, postmodern feminism presents itself as the privileged (if not only form) of social criticism, or to grass-roots struggles which might be more essentialist in outlook.

With Nzomo, Udayagiri wonders whether postmodern feminism is taking feminism in a direction that is too academic. This would obviously privilege feminist academics over local women activists. In her view, for (successfully) engaging in praxis feminism needs to include an element of essentialism.

The critiques by Nzomo and Udayagiri raise two issues that might take these debates in new directions. Although a good case can be made for the claim that political action, grounded in an "element of essentialism," is more transparent and possibly more effective in the short run, it does not necessarily deny the *possibility* of organizing on postmodern feminist principles. One could even argue that political action grounded in differences, and open to the limitations of both knowledge claims and action, could be more effective in the long run, both for its capacity to build alliances with oppositional groups and because it would (possibly) be more difficult for the state to repress or co-opt such decentered alliances. However, some problems connected with a postmodern feminist strategy for political action remain to be resolved. These include questions of how to select partners for building alliances and how to avoid possible divide-and-rule strategies by the state. Obviously, the entire question of postmodern feminist political action needs more discussions.

A second theme for debate is the possibility of rendering postmodern feminism less Western in its outlook, without losing its important insights on colonial/neo-colonial discourse and modernity. In her chapter, Barriteau Foster is heading in this direction. Taking the challenges and concerns of the contributors to this section as a starting-point, a non-Western postmodern feminist perspective would tend to be more praxis-oriented than current Western postmodern feminism.

In conclusion, the three authors reflect on quite a few of the themes – such as the question of political action, the knowledge/power nexus, the relation between identity, agency and diversity – that have been raised throughout the book. At this point postmodern feminism is more successful at raising issues and criticizing existing development theories and practices than formulating a transformative agenda with clearcut answers and solutions to women's developmental problems in the South and North. From the authors' comments it can be concluded that postmodern feminism needs to address a few issues before it can do more than raise questions within the field of Gender and Development. First, it needs to resolve the questions of how to engage in political practice/action. Second, it needs to consider ways to provide a basis on which to make "moral judgements," i.e. how to distinguish between "good" and "bad." Third, it needs to make sure that it will not become too elitist or academic. In Maria Nzomo's words: "postmodernism . . . must adopt itself to feminism, for the former to acquire significant and practical relevance to the woman question especially in the African context." The contributions in this section (and the book) are intended as a step in that direction.

7

WOMEN AND DEMOCRATIZATION STRUGGLES IN AFRICA

What relevance to postmodernist discourse?*

Maria Nzomo

INTRODUCTION

We want a world where basic needs become basic rights and where poverty and all forms of violence are eliminated. Each person will have the opportunity to develop her or his full potential and creativity, and women's values of nurturance and solidarity will characterize human relationships. In such a world women's reproductive roles will be redefined: child care will be shared by men, women and society as a whole – We want a world where all institutions are open to participatory democratic processes, where women share in determining priorities and decisions.

(Sen and Grown 1987: 80–1)

The vision for Third World women so well captured by DAWN in the mid-1980s, is one that many African peoples, especially women, clearly identify with in the 1990s. Following three decades of postcolonial autocracy and poverty, both internal and external pressures have finally led to the crumbling of the single-party regimes and the emergence of multiparty politics. For some countries, the long road towards democratization of politics and society has just begun. For others, civil strife and incumbent government resistance still make the situation uncertain. Nevertheless, the general mood among African citizenry is one of determination to replace dictatorial regimes with popularly and democratically elected governments. Popular struggles for democratization and development are everywhere the theme of conferences, seminars, workshops and even street demonstrations.

In Kenya, the repeal in December 1991 of Section 2A of the Kenyan

* Editors: Nzomo bases her critique on a rather narrow interpretation of postmodernist thought. Some of her points, such as the issue of unity in diversity, are now being voiced by postmodern authors (see Introduction/Conclusion and chapter 10).

131

constitution that paved the way for the return to multiparty politics opened the gates for popular participation in what is now viewed as the democratization process. Numerous interest and pressure groups have emerged, all demanding to have their interests and concerns included in the new democratic agenda.

Women more than any other interest group have come out very strongly demanding that their voices be heard: that their gender-based interests be included and mainstreamed in the new democratic agenda and that they participate on equal footing with men in the democratization process. Women activists and scholars have therefore embarked on a massive campaign of political mobilization, conscientization and sensitization of other women (and men) on the linkage between gender equity, democracy and development. In other words, the struggle against gender subordination is being linked with struggles against oppression based on national, class and other identities.

The experience in Kenya shows that African women, regardless of their class or ethnic background, do not need much convincing about their subordinate status and the need to seize the opportunity presented by multiparty democracy to change that status. The issues and demands that pervade the democratization discourse that is currently in motion among Kenyan women include the following:

1 That the universally accepted democratic ideals of democracy be adhered to in practice, especially in regard to equality of women and men;
2 that women be equal and effective participants with men at all levels of decision-making. The major issue therefore is to lobby, mobilize and strategize for the election and appointment of gender-sensitive women in large enough numbers in all policy-making and implementing bodies;
3 that laws and practices that discriminate and oppress women be changed, as they deny women their basic human rights on matters affecting, *inter alia*, family life, property ownership, employment terms and conditions and all forms of violence against women; and
4 that development strategies be fundamentally restructured to redress the existing feminization of poverty and the unfair division of labor that places increasingly heavier burdens of production and reproduction of society on women.

POSTMODERN DISCOURSE AND THE ASPIRATIONS OF AFRICAN WOMEN

Where does postmodernist discourse fit into these aspirations and hopes of African women in Kenya? Does it have any relevance? From my position as an activist involved in the Kenyan women's and democratization movements, the relevance of any theoretical approach or perspective first and foremost hinges upon its practical utility in providing viable guidelines in the search for answers to practical problems of human existence and development. The concepts, ideas and arguments advanced must then have practical applicability to *actual* rather than *abstract* situations.

Although many of the arguments raised by the postmodernist approach may be relevant to the experiences of women in Western industrialized societies, I am of the view that they do not offer immediate practical utility for women in Africa at the present time, as they struggle for the enhancement of their status in the changed context of a post-Cold War environment and economic hardships. However, the postmodernist critique of modernization theories and subsequent development theories that have dominated development planning and analyses for Africa, is to a large extent relevant to the analysis of the situation of African women, as it highlights the contribution of these theories to the increased marginalization of women in the development processes.

But before discussing these two arguments further, it is important to point out that my analysis is based on the following (rather simplified) interpretation of the postmodernist perspective:

1 There is no such thing as a universal reality or explanation of societal problems. We must therefore abandon the search for universals and adopt a more pragmatic, *ad hoc*, contextualized and historically specific analysis. We must acknowledge difference while avoiding universalizing essentialism.
2 Even concepts such as *knowledge, justice* and *beauty* do not have a universally acceptable criteria or meaning.
3 Postmodernism therefore emphasizes analysis that focuses on the individual "self" and *difference*, i.e. separateness rather than unitary themes of *solidarity, coherence*, etc.
4 Postmodernism rejects the universal applicability of modernist theoretical approaches such as Marxism and liberalism, arguing that they are associated with the political and conceptual baggage of their era.
5 Postmodernism also questions the modernist view that *modernization* and *progress* are obtainable goals (Nicholson 1990).

Given these postulates, this chapter will illustrate why the postmodernist approach is not in harmony with the democratic ideals and strategies guiding women in their pursuit of gender-sensitive democratization in Kenya. It will also examine the relevance of postmodernist critiques of modernization theory.

THE CONCEPT OF DEMOCRACY

Within the postmodernist approach, the concept of *democracy*, and the principles that hinge upon it, namely *social justice, freedom* and *equality*, cannot be treated as universal *ideals* or principles, but should instead be defined and applied only in the specific historic and social context in which the concept is being used. In other words, if for example, the Kenyan society wants to pursue democracy, then that democracy should be a Kenyan homegrown version of democracy, which does not attempt to derive its legitimacy or knowledge from a universal definition of democracy. Thus, from a postmodernist perspective, knowledge about how to create and sustain a democratic society should not be guided by existing knowledge on strategies and methods, but rather derived from Kenya's own experiences and local circumstances.

This is where the question of the relevance of postmodernist thinking becomes an issue in the African context. For while acknowledging that there are historical and socio-cultural differences that distinguish Kenyan society, it is difficult to justify a dismissal of the basic principles of democracy as irrelevant to Kenya, simply because they were developed in ancient Greece by political philosophers who knew nothing about Kenya. From past (failed) experiences with imported democracy during African decolonization, one recognizes the need to adapt democratic principles developed elsewhere to the specificities of the Kenyan situation. However, taking postmodern thought to its extremes would involve rejecting the outlines of democratic thinking and practice as they have been developed in other parts of the world. That is clearly counterproductive and naive, because it would require the democratization movement in Kenya, with its limited resources, to reinvent (rather than simply critique) established thinking about democratic practice and strategies of democratization.

For women particularly, their case for gender-sensitive democratization depends to a large extent upon their ability to convince male-dominated society that women's demands are well within the basic ideals of democracy and that these principles find support in the constitutions of democracies all over the world, in the Universal Declaration of Human Rights (1948), the Covenant on Human Rights (1976) and the Convention on the Elimination of All Forms of Discrimination against Women (1979). These are for the most part universally accepted legal standards of equality for women and men, which women in Kenya cherish as they struggle for equitable and effective participation and elimination of all forms of discrimination against them in the current democratization process. To remove the possibility for appealing to universal ideals would seriously diminish the strategies available to women for improving their position in society.

Postmodernist thought, as I understand it, does not have much to offer to the feminist movement now sweeping through Kenyan society. Women of Kenya feel particularly empowered to struggle for their basic human rights as women because they are keenly aware that the principles on which their struggles are based have universal backing and support. They are encouraged by the knowledge that their government has ratified the key international human rights conventions and hence can be called upon to honor its declared commitments. They are also encouraged by the solidarity that they believe they can expect to receive from the international community and other women around the world who also are experiencing various forms of gender-based discrimination. The postmodernist critique in this respect is not only irrelevant, but also if imposed on the current struggles for gender based democratization by African women, it could demoralize the emerging feminist movement and weaken the struggle rather than strengthen it. The danger, in other words, might be that this postmodernist critique will be used to undermine the Kenyan women's and democratization movements.

To further illustrate this point I will examine a key issue which the movement of women in Kenya is now contesting. This is the need for increased and effective participation of women in key political and public decision-making positions (Nzomo 1992). In the short term, women are mobilizing and strategizing to ensure in the near future women will constitute a *critical mass* of at least 30–35 percent of the total civic legislative bodies. One basic strategy to achieve this goal is to sensitize and conscientize women, who are the majority of voters, on the power of the vote and the merits of casting their votes for committed women rather than for men. The other related strategy is for women's political pressure groups to encourage and build confidence in those women with the necessary political will and commitment to contest political office in civic and parliamentary elections. Since the return of multiparty politics, nonpartisan women's pressure groups have been engaged in a process of capacity-building for women voters and women candidates with a view to increase the latter's chances of being elected into political office, and to enhance general civic and gender awareness among women.

Taken to its extreme, the postmodernist approach would be at loggerheads with the approach the Kenyan women are employing in their search for individual and collective empowerment. Postmodernism would argue that it is wrong to assume that women will, for example, vote as a bloc for other women. The assumption of solidarity and unity among women, postmodernism would argue, contains the danger of suppressing voices of those women who may have different views from those spearheading the dominant feminist democratization movement. This is a real danger, but the Kenyan women leading the feminist movement for democratization would argue that they have no intention of suppressing dissenting voices, whether from women or men. Indeed, as a demonstration of the desire to encourage dialogue and sharing of diverse experiences and to create individual spaces for women's divergent agendas, two national women's conventions were arranged in February 1992 and March 1993, which brought together Kenyan women from the grassroots to the national level in order to deliberate on their shared vision of the women's agenda in the current democratization process. Since the majority of women spoke with one voice at these conventions, this was taken as an indication of *unity in diversity.* In other words, despite their class, ethnic religious and other differences, women agreed to converge around their common subordination as women and to work together for their collective political empowerment. This however does not mean that there are no areas of competition and conflict between women as individuals and within and between groups.

A high premium has been placed on political empowerment as a means to achieving other goals associated with the advancement of the status of women. The argument has been that if women are in key decision-making and policy-making capacities in large enough numbers, they would, for example, exert decisive influence to ensure the removal or repeal of laws that discriminate against women and they would participate in designing development policies that mainstream rather than marginalize and disempower women. While bringing more women into politics would not *ensure* such woman-centered policies, legislatures without women will surely never bring about such changes.

Thus, while postmodernist discourse would emphasize *difference* and *diversity* among women, African feminists are emphasizing *unity in diversity* as a necessary strategy for strengthening the women's movement, their solidarity and their empowerment.

GENDER AS A SOCIAL CATEGORY

Postmodernist discourse quite rightly points out that women are not a homogeneous category, but rather belong to diverse socio-economic groupings based on class, ethnic or racial identities. Gender subordination is therefore not uniformly experienced by all women in the same way and with the same intensity at all times.

As noted earlier, women's struggles for gender-based democratization in Kenya, as elsewhere in Africa, depend to a large extent on group solidarity. Indeed, even before the emergence of the current movement towards democratization, Third World women have had a long history of attempting to empower themselves through women's organizations.

Thus while class and other forms of social identities are important, and can indeed limit the scope of participatory action, it is clear that when African women have broken through their socio-economic distinctions and spoken in unison, they have become a power to reckon with. Recent events in Kenya illustrate this point. In Kenya, women are constantly reminded that, being 52 percent of the population, power lies in numbers. And they have indeed on several occasions demonstrated their commitment to collective action in the current democratization struggles. A good example is the mothers and relatives of political prisoners who went on a prolonged hunger strike for a month in 1992, demanding the release of their sons who were political prisoners (Kiraitu 1992). The strike, led by elderly (one was then 83 years old) illiterate, rural women, received support from all Kenyan women (and some men) regardless of age, ethnic and class identities. The women's demands were based on their role as mothers and relatives of the imprisoned men. The hunger-striking mothers came to represent the vision and a viable strategy in the struggle for democracy in Kenya. Despite police brutality and repression, these women's continued defiance and determination to have their demands met, helped to strengthen the women's movement and its sense of empowerment. The similarities in strategies employed by these Kenyan women and those of the Argentinian mothers of the Plaza de Mayo are striking. Both groups involved ("traditional") images of mothering, i.e. caring for the well-being of the family, to oppose authoritarian regimes.

Postmodernism seems to have little relevance here, as it would dismiss or downplay the feminist solidarity that is developing in Kenya, which is challenging and uprooting organization and resistance based on gender-based oppression, as well as on national, class, racial and ethnic identities (see chapter 8 and chapter 9 for a contrasting view).

THE FEMINIZATION OF POVERTY

I agree with the postmodernist argument that the habit of national/ international development planners and aid agencies of lumping all Third World women together as one category, i.e. *poor* and *vulnerable*, is misleading and patronizing. However, it is important here to recognize and adopt the DAWN perspective, which notes that since the poor are the majority in the Third World and since women constitute the majority of the poor, development policy and planning that targets the poor would invariably impact positively on all women and society at large. Thus a development focus on poor women is quite appropriate, as long as it does not fall into the trap of overt patronizing and essentialism (see chapter 2).

In the current struggle for democratization in Africa, poverty among women is quite rightly being treated as a central concern. The economic crisis of the 1980s and 1990s has thrown even the few middle-class women into the category of the poor. Feminization of poverty has worsened as the implementation of the World Bank's structural adjustment programs have undercut many of the advances made by Third World women in the 1970s and have increasingly thrown women into the ranks of the poor. Moreover, it is important to remember that many women who have acquired middle-class status by marrying wealthy husbands, have no control over such wealth and hence are only well off as long as they remain married. Since they are thus *potentially poor*, they find themselves in a particularly cruel and vulnerable class position.

At the national level Kenya, like other African countries, is poor by any standards and will continue to be so unless and until certain internal and external restructuring is carried out. To that extent then, the concept of mass poverty must be employed to underscore the general situation prevailing in the country, at the national, gender and class levels.

Thus in the struggle for democratization of society, the political economy of female poverty is central to women's struggles. Poverty among women is linked directly to the question of women's economic empowerment, under which the following issues are addressed in the context of democratization:

1 property relations between women and men;
2 credit facilities, technical inputs and support services for women in production and reproduction; and
3 an enabling environment for women's advancement in professions and careers.

These concerns clearly cut across class lines and, consequently, women in Kenya prefer to address them in the more universalizing context of gender subordination and the feminization of poverty, rather than as a class or any other group. In their search for a way out of the feminization of poverty, women in Kenya are now challenging patriarchy and in this context are particularly insistent about the need to fundamentally restructure existing laws, both common and customary, as these laws to a large extent legitimize and reinforce women's economic disempowerment. Kenyan women's groups are also calling for better knowledge of the law, and support for women who are challenging laws and socio-cultural structures and beliefs that subordinate women. In sum, postmodernist thought is not particularly strong in analyzing and recognizing the totalizing effects of global political and economic restructuring on the feminization of poverty in Kenya, much less in suggesting adequate practical strategies to improve the situation of women in Kenya (see chapter 9).

WOMEN'S MARGINALIZATION IN DEVELOPMENT PLANNING AND ANALYSIS

The postmodernist critique is certainly correct when it points out that modernization theories, as well as dependency and neo-Marxist perspectives, which have guided development planning and scholarly analysis on development since the 1960s, have not treated women as an important issue. It was wrongly assumed that women's status, experiences and problems in the development process were the same as those of men and hence did not require gender-based planning or analysis.

The development crisis we see in Africa at the beginning of the 1990s is to a large extent a reflection of decades of experimentation with theories and models of development which were manufactured in the North and unsuccessfully tested in Africa and other Third World countries. The trickle-down Rostowian growth model of the late 1950s and 1960s (Rostow 1960), was replaced by the basic human needs approach. Even alternatives to these models, such as the dependency and neo-Marxist models, relied heavily on Western paradigms. By the early 1980s, none of these "scientific" models had borne fruit, as Africa sank deeper into an economic crisis characterized by, among other things, high levels of external indebtedness. The IMF and World Bank then took over control of the direction of development policies of African countries by imposing the infamous structural adjustment policies as a condition for further development assistance.

In all of these changes and policy shifts, African women have never been given adequate attention as the major producers and reproducers of labor and national wealth. For a long time, African women as a subject of research constituted a marginalized discourse largely undertaken by social anthropologists who relegated women to the private domestic spheres of marriage, household production and reproduction. These "scholars" romanticized, underestimated and ignored a wide spectrum of vital roles played by women in the public and private sphere of their societies. Indeed it was not until the pioneering study of Ester Boserup (1970) that the role of women in development received serious attention. Her work highlighted the serious consequences of women's marginalization within development policies, both for women and for economic development.

Some progress has been made since then, as exemplified by the emergence of Women in Development, and later WAD and GAD, as legitimate fields of study. The UN Decade for Women heralded the increased importance accorded to research on gender issues by development planners and international development agencies. But as Parpart has noted, "WID policy remained squarely within the modernization paradigm" so that "while development planners called

139

for better conditions for women, most development plans ignored the need for fundamental social change in gender relations and the possibility that women might organize to fight for this" (Parpart 1993).

In the meantime, the intellectual divergence of Third World women from those in the North had already become evident by the mid-1970s. This divergence of North–South interests in research on women was first registered in the 1975 Mexico World Conference where women from the developing world took exception and repudiated what they saw as the patronizing attitudes and intellectual imperialism of women from the North (see chapter 3). The same sentiments were echoed the following year by African women researchers at a conference held at Wellesley College in the United States (Papanek 1986; Wellesley Editorial Committee 1977).

The creation in 1977 of the Association of African Women for Research and Development Action (AAWORD) was an expression of African women scholars' desire to articulate the African women's reality from an African perspective capable of yielding action-oriented policy guidelines that would bring about positive changes in the lives of African women and the continent as a whole. AAWORD (1986) then published some research papers outlining the type of methodology that could yield relevant data on and for African women.

The DAWN perspective spelled out in the 1987 study by Gita Sen and Caren Grown is yet another manifestation of Third World women's desire to define and explain their own experiences, their hopes and their aspirations from their own point of view, as subjects rather than objects of study by outsiders (see chapter 2). Postmodernist discourse is quite in harmony here with the vision of African and Third World women generally in that it underscores the fact that development planners need to pay more attention to the concrete realities of Third World women's lives. They need to discover the real as opposed to the assumed goals and aspirations of these women, and to seek out indigenous women's knowledge as a basis for their policy formulation and practice.

To the extent that postmodernist discourse is critical of the modernist underpinnings of many development theories and their failure to adequately address gender issues in Third World development, it does indeed have some relevance to Third World women's experiences and analysis.

CONCLUSION

From the above analysis of the postmodernist perspective, and notwithstanding the relevance of some of its aspects as highlighted above, what strikes one most is that the perspective seems to question the very fundamentals of feminism itself. And yet, the feminist perspective is in my view quite vital if women's movements are to survive and prosper anywhere in the world. The postmodernist critique would indeed dismiss the current strategies and visions of African women whose struggles for gender-sensitive democratization hinge upon universalist feminist ideals. These ideals are manifested in the growing political consciousness among African women, which is leading to a strong sense of self-awareness, self-esteem, female solidarity and the questioning and challenging of gender inequalities in the existing social systems and institutions.

It seems to me then that it is postmodernism that needs to adapt itself to feminism and Third World conditions/knowledge if the former is to acquire significant and practical relevance for women, especially in the African context.

8

POSTMODERNIST FEMINIST THEORIZING AND DEVELOPMENT POLICY AND PRACTICE IN THE ANGLOPHONE CARIBBEAN

The Barbados case[1]

Eudine Barriteau

INTRODUCTION

Social science research on women in the Anglophone Caribbean[2] represents a significant development for feminist theory-building.[3] This body of work has established the multiple realities (particularly around race, class and gender) shaping the lives of Caribbean women. It has initiated the groundwork for critiquing the gendered epistemologies derived uncritically from Enlightenment political discourses which have informed so much Caribbean social science research. For research on Caribbean women to transcend the mere addition of women to the literature, however, the need to expose the gendered nature of Enlightenment theories becomes in itself both an epistemological and a political project. Without confronting and deconstructing these theories, and the approaches to development embedded in them, research on women will produce findings which do not challenge the concept and practice of patriarchy, and can often unwittingly reinforce it.

In the Anglophone Caribbean, postwar development policy was heavily influenced by neo-classical economic theory. Caribbean Nobel Laureate, Sir Arthur Lewis, advocated heavy reliance on foreign investment, export-oriented industrialization (Lewis 1966; 1978), and the creation of export enclaves requiring cheap (i.e. mainly female) labor (Kelly 1987; Ward 1986, 1990). The gendered nature of the neo-classical modernization paradigm produced gendered relations in development policy, planning and practices, which have stressed economic models as if these originate in value free, ideologically neutral development paradigms.

Recent feminist theorizing has disclosed the geopolitical, Eurocentric, masculinist, politics of development, and the specific ways economic development policies have produced differing outcomes for women (Momsen 1993; Stamp 1989). The emphasis on gender relations within development policy has clarified women's central position in development. Generally, however, feminist theorizing in the Caribbean has been embedded in the prevailing concept of a monolithic, homogenous, Caribbean woman. Indeed, both the liberal feminist integration model, and the socialist feminist exploited workers model, have fostered this tendency by focusing on low-income earners or women as wage workers. This has obscured the effects of gendered power relations on differing constituents of women. New feminist theorizing is needed to deconstruct and destabilize this construct and to incorporate gendered power relations into feminist analysis.

While some scholars have begun to address these issues (Momsen 1993; Safa 1992), the lack of a direct critique of the gendered theoretical frames originating within the context of the Enlightenment project has created a certain tension in the work of some Caribbean feminists. Many Caribbean scholars have articulated quite clearly that their experiences differ from those of North American feminists (Hart 1989: 5), yet they often use the language of European/North American liberal feminist discourse, particularly when speaking of the need for equality and emancipation. To speak of equality is to operate within a particular theoretical frame – that of liberal political theory. But this theory is derived from philosophical constructs which ignore women (Landes 1988; Nye 1988; Pateman 1980a and b; 1989). The inclusion of women has merely stretched liberal and Marxist discourses without challenging their patriarchal assumptions.[4] They cannot accommodate feminist interests which threaten their very foundations. Feminists will have to theorize outside established paradigms if they are to come up with theories that truly explain women's realities. This chapter explores the possibility that postmodern feminism may offer fresh ways of thinking about women's lives. The focus is on the Anglophone Caribbean, especially Barbados, but it has implications for the region as well.

DEVELOPING A POSTMODERNIST FEMINIST PERSPECTIVE

A postmodernist feminist approach to research on Anglophone Caribbean women should distinguish between research on women and feminist research. Research on women, with its emphasis on statistics and descriptive profiles, is a necessary starting-point for feminist research. Postmodernist feminist research, however, starts from the premise that "where there is society there is gender" (Smith 1987b: 4; 1987a) It seeks

143

to expose the gendered reality of social relations, particularly the influence of patriarchy on women's lives. It seeks to reveal and deconstruct gendered assumptions of the dominant discourses, and strives to produce new generalizations about the lives of women that are inclusive, guarantee equal access to power and recognize the diversity of women. It sees women as gendered, interacting in a social-cultural environment where class, race and sexual identity are social realities that affect women differently at different times. Finally it seeks ways to empower marginalized peoples through political action (Jones and Jonasdottir 1988: 27).[5]

Postmodernist feminist theorizing can change how development policy is formulated and implemented in the Caribbean. Opening up the concept of Caribbean women to include constituents other than working-class women does not mean abandoning the legitimate and pressing problems of low-income women. Instead it enables development planners to recognize that class relations alone do not produce exploitation. Deconstructing the imagery of the Caribbean woman as working-class victim requires that we do not make a virtue out of oppression. Postmodernist feminist theorizing does not deny oppression or reify it as an essentialist construct of a particular class or women. Rather, it reveals the multiple contested locations of relations of domination in women's lives.

The experiences of female entrepreneurs in Barbados provide a good example of the limitations of liberal and socialist feminist theorizing, as well as the promise of a postmodern feminist approach. Female entrepreneurs in Barbados (i.e. women who run business ventures, and own and control their means of production) do not fit the stereotype of the modal Caribbean woman. Their economic activities pose particular problems for liberal and socialist feminist theorizing. Liberal feminists emphasize the homogeneity of women and call for their integration into public life, which they see as gender neutral and therefore potentially beneficial to women. They see women entrepreneurs as potential leaders in this process (Goetz 1988: 418). To socialist feminists, female entrepreneurs appear as a privileged group of women, who need little help from development specialists. Thus, both liberal and socialist feminists believe female entrepreneurs are largely excluded from gender discrimination. Yet this is not so. A study of thirty-two women who started, own or manage their own businesses in Barbados reveals many misconceptions about the economic agency of these women. The study highlights several instances of discrimination arising from unequal gender relations. Many of the activities of these women are devalued and their business decisions assessed on the basis of androcentric criteria (Barriteau Foster 1993b).

These women are not microentrepreneurs and their business activities are easily measured. They satisfy the key criteria for inclusion in the formal sector (Bromley 1978). They deal with commercial banks, own separate business establishments, pay national insurance contributions for workers, advertise their goods and services and are registered with the Division of Corporate Affairs. Yet female entrepreneurs and the challenges they face are absent in the research on women in developing countries. They are excluded because they differ from the modal, low-income woman. As the owners of productive capital, it is assumed they should not face problems as women. Considerations of class relations are allowed to supersede investigation of gender relations. In contrast, a postmodernist feminist understanding of the contingency of the construct, the "Caribbean Woman," acknowledges the female entrepreneur as another category of "woman" subject to and resisting differing kinds of subordination.

The history of research on Caribbean women highlights the need for a locally grounded postmodern feminist approach. The first social science studies on Anglophone Caribbean women were conducted in the context of research on the family and kinship patterns rather than women as agents (Clarke 1957; Smith 1965), although some studies of women in the Hispanic Caribbean were approached through a literary lens (Courteau 1974; Hoberman 1974; Malachlan 1974; Pescatello 1974). But in the Anglophone Caribbean, women did not become the central focus of analysis until the late 1970s,[6] when the Women in the Caribbean Project (WICP) (1979–82) undertook the most extensive in-depth study of women in the Caribbean.[7] This research analyzed the multiplicity of activities that constitute work in women's lives, and described women's resourcefulness in developing survival strategies under adverse economic conditions (Massiah 1982a, b; 1986a; Barrow 1983; 1986a). However, it did not offer a Caribbean-based theoretical framework for analyzing these strategies and experiences. It has remained bound to traditional liberal, epistemological and methodological frames of analysis, which are gendered and cannot make sense of the complexities of Caribbean women's realities. It exposes the exclusion of Caribbean women from post-independence economic and political developments, but fails to critique the epistemologies and methodologies that maintained that exclusion in the first place.

The WICP research has also ignored the discourse on development and the way it has represented and reconstituted the lives of Caribbean women. It has missed the opportunity to interrogate the "Caribbeanization" of the women-in-development discourse, which has represented women in ways that bear no relation to their experiences. For example the 1988–93 development plan of Barbados addresses women primarily in the context of the nuclear family (Barbados 1989a). Yet this configuration is a myth, reflecting neither the history nor the culture of families in Barbados, which, whatever their socio-economic class, racial and ethnic identity, survive and flourish because of extended family networks rather than nuclear structures. The wealthier the families the stronger and broader the economic and social linkages of kinship. Moreover, in the 1970s, Caribbean development plans reflected WID discourse, including support for integrating women into development. Yet women have long played a key role in the economic activities of the region (Momsen 1993; Reddock 1988). Indeed, recent structural adjustment policies have placed even greater burdens on women in Barbados and elsewhere in the region (Barriteau Foster 1993a). However, the inadequacy of existing feminist theories to help us make sense of this condition of material and psychological decline has not received similar attention.

Ironically, the research and activism of various Caribbean groups and scholars[8] substantiates the multiple dimensions of women's lives and the need for a new approach to the study of Caribbean women. The findings complement, contradict, compound and confound existing stereotypes of Anglophone Caribbean women. Women emerge as economically vulnerable and insecure (Barrow 1986a; Powell 1986); display alarming levels of female self-contempt (Clarke 1986); doubt their abilities to be effective leaders (Clarke 1986); and defer decision-making to their male partners (Odie-Ali 1986). They define their sexual identities and roles as an intense commitment to mothering (Powell 1986); and "They recognize they must accept male domination and a male dependent role" (Anderson 1986: 311). Simultaneously these women are resourceful, decisive and self assertive (Barrow 1986b; Clarke 1986; Durant-Gonzales 1982). They dominate decision-making in the household (Powell 1986) and 82–92 percent of them said gender made no difference to their personal development (Clark 1986). Sexually, they adopted a practical attitude towards prostitution as "work" (Anderson 1986: 300); and while they desired relationships with men based on reciprocal emotional support, companionship and economic support (Massiah 1986b), they felt it might be necessary at times to have extra affairs with other "women's partners" to obtain money (Barrow 1986a: 168). They have a strong sense of their equality with men and they strive for independence or at least interdependence in their relationships with men (Anderson 1986: 311).

This information has been viewed as contradictory (Anderson 1986: 316, 318) rather than as illustrating the multiple, shifting interactions of women's lives. Attempts have been made to apply existing social science theories to this complex picture. While WICP researchers seem aware of the limitations of this approach, they have not yet attributed them to inadequate theoretical frames. Christine Barrow hints at this paradox when she notes, "Development planners and researchers totally misconceive the Caribbean situation when they assume that women are confined to the domestic domain, are not aware of the world they live in and cannot articulate their perceptions, values and problems or actively devise strategies for their own welfare and that of their dependents" (Barrow 1986a: 170). She rejects the cultural dualism model which defines Caribbean women along two axes: Afro-centered economic autonomy and Euro-centered passivity (Barrow 1986a: 165, 169).[9]

In an effort to move towards a more nuanced and relevant approach, the following construct is offered as a step towards building on the multiple, messy, lived diversities of Caribbean women's lives. It draws on the specificities of Barbadian women's experiences, but has wider implications as well.

ASSUMPTIONS OF THE NEW CONSTRUCT

Social science research that continues to rely on epistemologies which maintain the exclusion of women are reinforcing dominating relations of gender. Unless we deconstruct these received knowledges and frames, dominating relations of gender will continue to permeate social science research in the Caribbean. However, a new theoretical construct cannot simply insist that Anglo-Caribbean women's perspectives are truer, clearer or more meaningful because they live the particular confluence of their history and culture. This would deny the relevance of other women's voices and experiences. Three considerations must inform the building of this new theory/construct: recognizing difference; organizing political action based on both differences and commonalities; and acknowledging the gendered nature of all social relations.

Some clues for this undertaking have been supplied by the WICP researchers, who recognized the need for relevant, culture-specific feminist theorizing that would investigate "gender ideology and explore the ways in which cultural concepts of gender guide social relations in the Caribbean" (Anderson 1986: 320). Massiah points to the difficulties posed by research on gender relations when both those being analyzed and those conducting the analysis are affected by the same gender relations (Massiah 1986a: 176). Anderson advises researchers to view gender relations "as a distinct subsystem in the same way as economic or political subsystems" (Anderson 1986: 321).

Caribbean society shares with all societies the gendering of social relations, even though gender manifests itself differently in different cultures. The new theory/construct sees Caribbean women and men as equally gendered and is willing to explore the diversities made visible by this approach. It rejects the definition of woman as not-man, arguing that what we know as "female" and "male" are equally learned behavior – there are no essential feminine and masculine traits.

The WICP findings testify to the dominating relations of gender in

Caribbean society. Yet they reveal important instances where both women and men have tried to work around this pervasive construct. Women act continuously in their daily lives to resist and overcome gendered relations. Men sometimes attempt to resist gendered constructs as well by refusing to accept stereotypes of women (Barrow 1986b: 61). This behavior illumines the often overlooked fact that men are also victims of this construct, even though gendered relations overwhelmingly operate in their favor. Recognizing that both women and men are socially recreated by gender shifts the analytical frame from women as victims of a patriarchal system to a view where both women and men are equally implicated as the subjects of gendered constructs. Both are located within gendered societies. However, the recognition that men are equally gendered does not alter the reality that all social relations are asymmetrical and usually benefit men.

This reconceptualization of gender necessitates a different methodological approach. It requires investigation into the way historical, cultural, social and economic issues have impacted differently on women and men. It rejects identical conceptual frames for women and men. It sometimes requires researchers to focus simultaneously on women and men, but it always maintains gender-specific distinctions. Above all, this new construct pays attention to the interactive experiences of racism, colonialism, and the legacy of being viewed as inferior subjugated "others." It examines and deconstructs the Enlightenment discourses which have marginalized Caribbean women and condoned slavery, domination and subjugation, while recognizing the multiple jeopardies confronting Caribbean women, the fluidity and ever-changing dynamics of their lives, and the need for political action to correct asymmetrical gendered relations.

THE NEW CONSTRUCT/THEORY PRESENTED

The emerging construct for studying Anglophone Caribbean women, particularly in Barbados, attempts to distance itself from all universalizing, hegemonic tendencies of other theories. It is grounded in Anglophone Caribbean women's experiences rather than imposing a predesigned package of assumptions and generalizations to be tested upon women. Thus far we have discussed the contours of the theory; the space within is filled by operationalizing gender, class, race, sexual identity and political action in the Caribbean context. The proposed construct/theory views all past generalizations on Anglophone Caribbean women as "subject to scrutiny and change."

As we have seen, previous research has for the most part approached women as both non-male and a subset of man. It has generally assumed that women's realities can be understood or revealed by employing androcentric frames of analysis. The proposed construct/theory challenges this assumption and calls for research informed by postmodern feminist gender analysis. It draws on Jane Flax's analysis where gender is both a thought construct that helps us understand the environment in which women live, and a central social relation that shapes all other social relations and activities (Flax 1989: 60). Woman is distinguished not by being non-male, but by the social construct of gender and the particular way society interacts with her. Questions can no longer be posed as woman in relation to man, rather one must examine gendered woman interacting with and being acted upon by her environment. The frame has shifted.

For example in examining women's political participation, the social construct of gendered behavior reveals that women often react to "politics" in ways that society has socialized them and expects them to act. This approach is also concerned with how women perceive that interaction, and finally how they react to society's expectation – either through conformity or rebellion, or by redefining "politics." When women refuse to participate in politics which exclude their interests, mainstream scholars define this rebellion as non-participation. Yet women are in fact acting politically; they are rejecting politics that excludes their concerns. The new theory/construct is not concerned with whether women participate in elections as men do, or why women are perceived as less interested in politics than men. That approach explains women in relation to men, and retains a liberal political definition of political (or apolitical) behavior.

Instead the new theoretical frame shifts its focus to gendered woman and asks, what do Caribbean women consider political? What factors within their environment do women consider more important than mainstream political activities? The theory does not assume that women should be interested in political participation as defined by men, nor does it assume that woman rank, or should rank political participation on a continuum of priorities. It allows women to redefine politics and political participation, and validates women's approach to the political process. Indeed, this emerging construct seeks to, "explore aspects of social relations that have been suppressed, unarticulated, or denied within dominant (male) viewpoints" (Flax 1989: 67).

This approach is guided by four principles:

It seeks to articulate feminist viewpoints within the social world [Caribbean environment] in which we live; it thinks about how we are affected by this world; it considers the ways in which how we think about them may be implicated in the existing power/

knowledge relationships; and it imagines ways in which the world ought to/can be transformed.

(Flax 1989: 71)

The gendered Caribbean woman interacting with her environment remains at the core.

Deborah King's analysis (1989: 75) permits us to build on the post-modernist feminist core of the theory. Theoretically and politically her contribution recognizes that much of feminist theory represents white, Eurocentric feminist theorizing and therefore fails to address the episte-mological and practical concerns of other women, especially black women. These theories need to be made more contextual. White feminists who continue to work within Western male-centered traditions should realize that their theories generally ignore the complex experi-ences and realities of women from other racial, ethnic and geographical backgrounds. This is damaging for a general feminist project because it perpetuates discourse based on exclusion and oppression.

Conceptualizing and building the concepts of race, class and sexual identity into this new theory encourages researchers to recognize differ-ences among women. It reveals the complex and shifting interaction between gendered relations, race, class and sexual identity. When women are no longer seen simply as "not-men", the futility of lumping them together is exposed. The inclusion of multiple identities fits well with the postmodernist feminist focus on diversity and difference. It changes one's questions and highlights the way cultural contexts shape women's lives.

This approach allows for differences among women of the same race but with differing historical experiences (for example, African-American and Afro-Caribbean women), for divisions along ethnic lines (Indo-Caribbean as opposed to Afro-Caribbean), as well as for differences among women with similar racial, historical, and cultural experiences but with different class backgrounds. It recognizes that these multiple identities overlap and interact in complex ways, changing at different times and in different social circumstances. For example a working-class Barbadian woman may either privilege being black, that is occupy race, or she may emphasize her working-class status, or she may emphasize a lesbian sexual identity. On the other hand, others may treat her in various ways, independent of her feelings about that interaction.[10] Particular identities may come to the fore in different situations. For example, when a middle-class Barbadian black woman speaks with her working-class black domestic helper, despite their common historical heritage, their class positions influence their interactions. However, they may both occupy the same position, or may be treated the same way in another social context on the basis of race. Yet in each interaction, a woman is still gendered, but her experience of gender varies with specific contexts and situations.

The WICP data supports the assumptions of this approach. For example, it highlights the differences between Afro-Caribbean women's experience of race in the Caribbean and that of black women in North America. Most Caribbean countries are predominantly black (if you include non-whites from outside the African diaspora). They range from 95.4 percent to 79.9 percent of the population (White 1986: 65). Blacks in the Caribbean face a legacy of race rather than the daily indignities faced by North American blacks. For example in Barbados, indigenous whites own or control economic activities within major industrial sectors, and maintain a web of corporate inter-linkages based on ethnic and kinship ties (see Beckles 1989b). Yet black Barbadians experience little racial discrimination in the areas of health services, education, transportation, housing and public sector employment. The proposed theory would acknowledge and analyze these differences rather than simply lumping blacks together as one homogenous group.

To achieve this flexibility, the concept of positionality must be incorporated into our analysis. Positionality states that the gendered subject can become imbued with race, class and sexual identity without being over-determined by them (Alcoff 1989: 318). Adapted to the Caribbean context, this emphasis on positionality recognizes the fluidity of women's lives, and the role of historical, economic, cultural and political institutions and ideologies of Caribbean society on women's lives. The Caribbean woman's identity results from how she interacts with, interprets and reconstructs her environment, and it may shift with changing circumstances (Alcoff 1989: 323). Thus positionality captures both the agency and contingency that postmodernist feminist theorizing offers to Caribbean women.[11]

Postmodernist feminist theorizing enables us to politicize these differing manifestations of gender subordination without imprisoning us in an essentialist, fixed frame. We can pursue coalition politics and undertake political action informed by a recognition that several factors impinge on the subjectivity of Caribbean women. This approach frees us constantly to relocate the site of resistance. Unlike Nzomo and Udagaryi, I believe the postmodern feminist focus on the fluid, changing nature of social relations does not have to inhibit political action. Instead it strengthens the many stances we can adopt. It enables us "to theorize the multiplicity of relations of subordination" (Mouffe 1992: 372) and to act on them. For example, female entrepreneurs in Barbados provide much needed employment as well as goods and services. Yet they are often misunderstood by government planners. As feminists do we ignore them because they appear privileged? Or do we recognize that irrespective of class relations, women experience gender subordination? Some argue that

Caribbean women are closing the gap between male and female labor force participation rates and that women are gaining more new jobs than men. Do we recognize this as an achievement or do we focus on the still generally poor conditions of women's labor?

Alcoff's concept of positionality and Mouffe's development of the strength of the plurality of various subject positions to feminist political agency, equip us to face these decisions and to practice a feminist politics free of an essentialist reading of the Caribbean woman and her problems. Recent issues which could be mobilized around include: the alleged dramatic increase in the rape of Indo-Trinidadian women;[12] the freedom of Barbadian women to attend male striptease shows; the impact of unprecedented levels of migration on Guyanese women and children. For any of these we can take political action deploying a postmodernist feminist concept of agency, that is, we can develop a politics based on the complex and fluid nature of women's lives rather than a fossilized, essentialist reading of race, class or social freedoms.

THE THEORY/CONSTRUCT APPLIED

The findings of WICP on women's political participation illustrate how different conclusions can be reached by utilizing this new theory. The WICP survey investigated women's participation in political life by examining voting patterns of women, noting the number of women represented on executive councils, the proportion of women represented on national assemblies and female presence on statutory boards (Duncan 1989). As we have seen, women's participation in male-dominated and defined political institutions are usually measured by both male and mainstream standards of political participation (Jones and Jonasdottir 1988: 22). The WICP data suggest that, "women do not actively participate in the political and policy making arenas of their societies" (Clarke 1986: 147). However, analysis with the new feminist theoretical construct in mind reveals that this research has adopted a purely male and mainstream criteria of politics and political participation. It never investigated whether political parties articulated a bundle of concerns that could be treated as women's interests (Duncan 1989). Women's race, class and gender were used to describe their location in a hierarchy of political participation, not to analyze their political behavior. Women's low activity rates were considered proof of their inadequacies rather than of sexist bias in the definition of "politics," or evidence of material obstacles to women's full participation (Jones and Jonasdottir 1988: 21).

But the women interviewed were vigorously involved in political activities, despite the constraints of operating within male-dominated political institutions. As we have seen, Caribbean societies are relatively racially homogeneous, so women's political activities have more often been pursued along class rather than racial lines (this may differ where there are large Indo-Caribbean populations) (White 1986: 65). In St. Vincent 75.9 percent of women in the sample were black, 1.7 white and 19.4 mixed. In Antigua 95.4 percent were black, 0.8 percent white and 3.2 mixed. The women's arm of the St. Vincent Labour Party ran an airport restaurant and plowed the profits back into the main party.[13] The Women's Action Group in Antigua managed an income-generating project in the agricultural sector. They used the profits to subsidize typing classes for underprivileged women in the group (Clarke 1986: 124). The members believed they were instrumental in the successful campaigning of their parties, for they had gone from house to house and thus established personal and political contacts within the communities. Clarke reports the women's own assessment of their effectiveness in political campaigning:

> Women have more stamina than men in campaigning, and they tend to be more enthusiastic.
> The women were very well organized; they really made an impact because they chat more, talk to more people.
> Women did the major part of the campaign.
>
> (Clarke 1986: 125)

These women are boasting of delivering the votes to their parties, of undertaking activities to benefit themselves and the party simultaneously. Only a mainstream, androcentric definition of politics would deny the political impact and relevance of these forms of participation. Furthermore women participate in "mainstream" political activity when it suits their interests. In Grenada, for example, women's involvement in organizations increased under the People's Revolutionary Government, which had instituted a comprehensive package of measures to benefit women and a policy to research the needs of women and to encourage their active participation in the development of Grenada (Joseph 1981: 17). Barriteau and Clarke found that 75.5 percent of a sample of the Grenada electorate believed the People's Revolutionary Government had improved social conditions affecting the lives of Grenadian women (Barriteau and Clarke 1989: 67) even though that same administration has been correctly described by Mies as a socialist patriarchal genealogy (Mies 1989: 204).

The construct can be illustrated by looking at two women's organizations. The Red Thread Collective of Guyana is a testimony to women's refusal to adopt the victim role. The Collective shows how women can create change when corrupt national bureaucracies and imperialist, international financial institutions fail people in the South. The Guyanese economy is in crisis, yet the women of the Red Thread Collective have fought back by producing creative projects to alleviate economic hardships for children, women and men in Guyana, including the production and sale of low-cost exercise and textbooks, and the manufacture of hand-embroidered greeting cards. In the words of one of its members,

> the daily issues are the issues of food and water and electricity, obviously we are talking of women being in the front line of the struggle in Guyana. And women on the front line of that struggle have found out who else is in the front line. By the very nature of women's work, the food line is where you find women.
>
> (Andaiye in Women and Development Studies Unit 1990: 33)

In Jamaica, the Sistren Theatre collective continues to blaze a trail undertaking a multiplicity of projects including drama, journalism and art. Sistren's method of information dissemination has been used by many women's organizations to explore and analyze the complex issues confronting women. For example, to investigate a particular problem, Sistren members spend days observing and participating in the activities of the women in question. They then write a documentary around this research and deal with several complex issues through dramatization and audience participation. The research work of CAFRA and WAND (on women and agriculture, law, micro businesses, youth and history) have sensitized even mainstream regional organizations such as the West India Commission and the Regional Council on Economic Development, which have requested input into regional policy formulation from these organizations.

These are examples of women's political participation that the theory explores and considers relevant. The activities of these new women's organizations fall outside the traditionally defined political participation, but they are all-important within their respective societies. These new organizations are not interested in the conventional definition of politics. They have decided that whatever issues confront women, whatever social interactions attempt to subordinate them is a political issue. They recognize the asymmetrical nature of social relations for women and attempt to overcome these. They refuse to await political change and instead actively seek to introduce it, infusing this change with a feminist perspective.

The proposed construct thus lets women (in all their complex, multiple identities) define the issues that should inform policy and practice. In contrast, mainstream approaches to women in development in the Caribbean have failed to connect widespread social, cultural and economic disruptions in women's lives to the shortcomings of existing development policy and practice. This has led to superficial analysis and irrelevant policies. For example, the Bureau of Women's Affairs in

Barbados, the government department responsible for issues concerning women and development, remains largely oblivious to the failure of development policies and the increasing burdens placed on women. The Bureau sees development as the creation of microenterprise, low-income projects. The 1983–8 development plan called for the Bureau to collaborate with various organisations to identify projects with income generating potential for women (Barbados 1988: 87). The Bureau has admonished women for not taking advantage of skills training courses and marked International Women's Day with yet another exhibition on women engaged in the manufacture of jams, jellies, pepper sauce and clothing.

In contrast, postmodernist feminist theorizing connects the crises experienced by young women, in fact all women, with the inherent rigidities of growth-oriented models. It detects the contradictions between prioritizing the values of consumerism, of mass consumption as advocated by the modernization paradigm, and the increasing pauperization and subordination of many women. Caribbean women and men are expected to consume more to fuel the economy, but the welfare state is shrinking. Certain services of education, welfare and health are returned to the private domain to be supplied by women's unpaid labor (Barbados 1989b), at great cost to their material and psychological well being. The gender implications of economic crises and its consequences are not examined. Yet Caribbean governments count on women to provide the services the state relinquishes. Indeed, productivity and efficiency are appearing in the public sector because costs have been shifted from the paid economy to the unpaid economy (Elson 1991), from governments to women in households.

By developing gender as an analytical tool within a postmodernist feminist perspective, one can transcend the dichotomy of private and public. This approach enables researchers to investigate the way conditions affecting women at one level have consequences at the other. This is particularly important in the critique of structural adjustment. In contrast, the WID discourse as practiced in the Caribbean ignores relations of gender and their consequences for women and still calls for the incorporation of women into development. A Gender and Development approach offers much more nuanced appreciation of gender relations and women's reproductive labor, but ignores the impact of race and culture. In contrast, a postmodernist feminist approach sensitizes researchers to the multiple realities of women's lives, thus avoiding an essentialist, modernist approach to women's problems. It calls for a different kind of development policy and practice, with differing outcomes for Caribbean women, depending on the complex, circumstances of their lives. It would link seemingly disparate issues on women to the social construct of gender, and expose the way gender ideologies are being manipulated to maintain the subordination of Caribbean women. Only then can one design political action that can transform women's lives.

CONCLUSION

As we have seen, postmodernist feminist theory explains the political action undertaken by many Anglo-Caribbean women[14] which liberal or socialist feminist frameworks have been incapable of addressing. While endorsing all activities that benefit women, this approach recognizes the limitations of incremental gains for women as long as they remain in societies with socially created unequal gendered relations. The proposed approach draws heavily on postmodern feminist insights. It emphasizes diversity, difference and positionality, while recognizing the importance (and power) of women's knowledge. It calls for greater attention to women's agency and recognizes that women and women's organizations are not merely adjuncts to male-dominated political structures. They arise from women's own agendas and are significant in themselves. The organizations discussed above are trying to influence policy-makers in order to improve their societies, and their actions should be taken seriously. But they are for the most part not interested in working for established political parties or governments. They are committed to women's empowerment, to the centrality of women's activities, and to the need to devise and promote new strategies for women while critically evaluating and discarding timeworn, irrelevant generalizations about Caribbean women.

This construct/theory also recognizes the long-established tradition of resistance among Caribbean women (Brereton 1988; Brodber 1982, 1986; Reddock 1985, 1986;). Beckles' *Natural Rebels: A Social History of Enslaved Black Women in Barbados* shows how slave women forged cohesive bonds, "in order to resist the divisive powers of race, colour, and class" (1989a: 175). The fluidities, contradictions and complexities of Caribbean women's lives, their actions to create space and resist dominating gendered relations existed during slavery as it does today. "Women fought, ran away, committed suicide and murder, became mistresses and prostitutes to their oppressors, terminated the lives of their infants as well as raised large, emotionally cohesive families . . . these actions and decisions suggest [women] sought to do the best for themselves in the only ways that seemed possible" (Beckles 1989a: 177). The remarkable similarities between the seventeenth- and eighteenth-century findings and twentieth-century research underscores the continuity of Caribbean women's quest for agency and control over their lives. This is a proud and complex history, one that the proposed theory celebrates, for it acknowledges women as actors, operating in multilayered material and discursive environments. Indeed, the new construct exposes the weaknesses of much current discourse on development for Caribbean women, particularly the assumption that Caribbean women have no solutions to developmental problems. However, this construct/theory is just a beginning. It calls on researchers and practitioners to adopt a more nuanced, situated approach to the study and practice of gender and development in the Caribbean. Above all, it challenges feminists in the region to build indigenous feminist theor(ies) which can both guide research and contribute to global feminist theory-building.

NOTES

1 This article has been extensively revised with assistance from the editors. An earlier version was published in *Social and Economic Studies* 41(2) June 1992, entitled "The Construct of a Postmodernist Feminist Theory for Caribbean Social Science Research."

2 While this theoretical frame should be applicable in different geographical and cultural settings, I deal specifically with the English-speaking Caribbean countries, because of their relatively homogenous historical and cultural background and from familiarity with the research in this area. Indeed the chapter focuses predominantly on Barbados. The term Caribbean is used to refer to the English-speaking independent Caribbean countries and the British dependencies.

3 I refer to the published material on Caribbean women in *Social and Economic Studies* 35(2 & 3) (Massiah 1986a, b and c); the seven-volume monograph series of the Women in the Caribbean Project (WICP), Institute for Social and Economic Research (ISER), EC, University of the West Indies; Ellis 1986; Hart 1989; Massiah 1988; Mohammed and Shepherd 1988; Reddock 1988; Safa 1986; and Young 1988.

4 The term feminist is used deliberately and precisely and needs no qualifiers. For more on the androcentric and misogynist biases in mainstream Western political theory, see Barrett 1980; 1985; Cocks 1989; Collins 1990; Delphy 1984; Di Stefano 1990b; Flax 1990a; Folbre 1991; Okin 1979; and Pateman 1989.

5 Although postmodernist feminists concur with postmodernists on an epistemological critique of Enlightenment theories, they disagree over the fundamental character of gender relations (for more, see Introduction/ Conclusion).

6 The period with little research on women coincided with the period of decolonization, independence and development. The New World School of scholarship with its focus on political economy flourished at that time in the Caribbean.

7 Besides the two-volume collection of fourteen research papers published in *Social and Economic Studies* (2 & 3, 1986) and the seven-volume monograph series the WICP established and maintains a computerized data bank on women in the Caribbean. It was also instrumental in establishing the first women studies program at the University of the West Indies.

8 The findings and conclusions of WICP indicate a need for a new conceptual framework (Anderson 1986: 314–21; Clarke 1986: 150), as does the research and activism of some of the new Caribbean women's groups such as the

Women's Forum of Barbados, the Sistren Theatre Collective of Jamaica, Working Women of Trinidad and Tobago, Red Thread Collective of Guyana and the Belize Rural Women's Association. The Caribbean Association of Feminists for Research and Action (CAFRA) and the community-based, feminist-informed development agenda of the Women and Development Unit of the University of the West Indies raise similar issues on a regional basis.

9 This model attributes patterns of behavior which are aggressive and seek autonomy to West African antecedents, and those that are submissive and conservative to Anglo-Protestant traditions. Behavior is associated with wealth. The rich are supposedly more passive while poor (working-class) women strive for economic autonomy. The project found both types of behavior in working-class women (see Barrow 1986).

10 Some find the concept of choice and subjective approach to a social interaction problematic. For example, they argue that blacks cannot define themselves in a racist society. This generalization does not hold. For example, in South Africa, whites have "objectively" determined that black-skinned South Africans are "black," that is politically and culturally inferior. But black South Africans have resisted that definition and chosen to be black the way they define it. This position challenged white definitions and established the possibility for counter-discourses on race/blackness.

11 This is particularly appropriate in the Caribbean with its complex, multicultural, multiracial history.

12 Rape is an ongoing issue for all Caribbean women. The alleged increase in the rape of Indo-Trinidadian women was politicized by a female Indian politician. It became a volatile issue intersecting gender, race and class relations.

13 Clarke however notes the restaurant was serviced by men. It would have been interesting to discover whether this was an example of the male party members exerting influence over the women to gain the contracts for this service or whether it was a sharing of the economic rewards of the venture.

14 This is not to suggest that these women's organizations describe themselves as postmodernist feminists, are interested in postmodernist feminism or even describe themselves as feminists. Some may subscribe to a socialist feminist framework without recognizing the disjunctures between their professed theoretical preference and their feminist praxis. The main point is of all the feminist theories available to explain the day-to-day realities of Caribbean women only the postmodernist feminist frame provides an adequate fit.

9

CHALLENGING MODERNIZATION

Gender and Development, postmodern feminism and activism

Mridula Udayagiri

INTRODUCTION

Over the last decade, new writing on Third World women has come to draw upon the theoretical perspectives of postmodernism and poststructuralism, particularly insights into the politics of representation and difference. These themes, which explicitly uncover essentialist and universalist interpretations, have become a paradigm for a new generation of feminist scholarship on Third World women, most notably in Mohanty's "Under Western Eyes: Feminist Scholarship and Colonial Discourses" (1991a)[1] and Ong's "Colonialism and Modernity: Feminist Re-presentations of Women in Non-Western Societies" (1988; see chapter 3). In the early 1990s, these two articles became mandatory reading for students of Third World studies. This chapter explores the implications of this new genre of scholarship for practical concerns that orient much of the theoretical content of the Gender and Development literature.

My interest in such writing evolves from my engagement in contemporary discussions on social and economic transformation in the South, which continue to be framed largely within "political economy" terms. As a person who entered North American academia with practical experience in development policy, my motivation to participate in the debate stems from the need to re-examine the significance of postmodernist analyses of discourses on women in the South. I am concerned with its relevance for understanding and creating change. Does it yield a blueprint for practical action in development as its protagonists claim? I do not claim to present an exhaustive review of postmodernism, nor of the critiques against it. Instead, I examine the *political* potential of contemporary postmodernist analyses of writing on women in the Third World and pose some provocative questions about the significance of such analyses for development policy and practice, especially for women.

Studies of change and transformation in the South have been described as studies of "development," "modernization," "underdevelopment"[2] and for gender-related issues has been associated with "Women in Development" (WID) or more recently "Gender and Development" (GAD). Suffice it to say these are political discourses, which legitimize mainstream definitions of modernization and development. As Marianne Marchand points out in chapter 3 "development" has been a problematic concept – not only because it accepts and perpetuates unequal relations in the global economy, but also because it ignores the possibility that people may perceive "progress" or "development" very differently from policy makers. For instance, the poor may desire greater material well-being without altering their patterns of social life – such as the Chipko movement (Shiva 1988). Development, then, is a contentious term, especially with the emergence of neo-liberalism in the post-Cold War era and the emphasis on market reforms as the principle mechanism to mediate contemporary core–periphery relations in the global system.[3] But even theories of development and underdevelopment remain firmly anchored in emancipatory paradigms that emerged in the age of Enlightenment.

Generally postmodernism breaks loose from social science's traditional philosophical concerns. It is no longer anchored by Enlightenment meta-theory, instead it is expected to be contextual, local, arbitrary and *ad hoc*. Contemporary postmodern analyses of discourses on women in the Third World, such as those by Mohanty and Ong, pick up on these themes, in particular drawing from Edward Said's *Orientalism* and Michel Foucault's insights into discourse, power and knowledge. They are effective theoretical approaches with which to challenge the essentialism and universalism that undergirds much of development and underdevelopment theory, especially as these are applied to women's experiences in the South (see also chapters 1 and 3).

Said's *Orientalism* (1978) was instrumental in showing how Western cultural perceptions of the Orient became a means of controlling the regions and peoples so classified. Academics and other specialists in the North have been able to control official knowledge of the Third World. Orientalism, or the knowledge of the Orient, has been created:

> for dealing with the Orient – dealing with it by making statements about it, authorizing views of it, describing it, by teaching it, settling it, ruling over it: in short, Orientalism as a Western style for dominating, restructuring, and having authority over the Orient.
>
> (Said 1978: 3)

Said's critique has been taken up by historians such as Prakash, who condemns Orientalism for:

First, its authoritative status; second, its fabrication of the Orient in terms of founding essences invulnerable to historical change and prior to their representation in knowledge; and third, its incestuous relationship with the Western exercise of power over what we call the Third World.

(Prakash 1990: 384)

In a similar vein, Mohanty and Ong deconstruct writing on Third World women to show how the same representations are reproduced in contemporary discourses in the postcolonial period. The concern is with "the production of the Third World Woman as a singular monolithic subject in Western feminist texts" (Mohanty 1991a: 51) wherein such an image is founded upon the presumption that Western women are truly emancipated from patriarchal shackles.

Correspondingly, Foucault's conceptualizations of power are of particular relevance to the fiction of essentialism. Foucault's theories of power have become a fecund source of postmodernist argument (Harvey 1989: 45) and have fueled a growing industry in critiques of modernist/neo-colonial discourses on the Third World (See Escobar 1984–5; Mohanty 1991a; Ong 1988). Universalist and essentialist categories, especially with regard to women in the South, theorize power in limited ways.

It is usually expressed in "binary structures – possessing power versus being powerless. Women are powerless unified groups" (Mohanty 1991a: 71). Since Foucault's theories recognize the manifold structures of power, with their varied forms and multiplicity "of localized resistance's and counteroffensives" (Escobar 1984–5: 381), they can show that women, far from being powerless, are agents in their own fates. Thus Foucault's perspectives on power provide a theoretical retreat from universalized conceptions of women in the Third World.

Ong's and Mohanty's inquiries into hegemonic "Western" feminisms are trenchant, powerful and effective cultural critiques of the subjugation of Third World women's experiences and identities in Western feminist discourses. However, they place an undue emphasis on universalist and essentialist categories, and fail to address their implications for political action within the context of change in the South. More in the spirit of eliciting a dialogue, rather than providing a definitive answer, this chapter explores the question – does postmodern analysis constitute the only genre of critique that can contribute to political action because it engages and confronts "difference"?

ANTI-ESSENTIALISM AND POSTCOLONIAL POLITICS

Both liberal and socialist feminist literature on women and development[4] supply the texts for the postmodernist analysis by Mohanty and

Ong. Like others drawn to this approach, language is the central focus of critical inquiry; they deal exclusively with textual study. This emphasis on textual analysis is part of a general shift that marks a particular historical development in American academic scholarship in the late twentieth century.[5] Unfortunately much of this scholarship denies the validity or political viability of other genres of social criticism. This privileging of postmodernism as a legitimate form of social criticism disturbs Bryan Palmer who argues that the prevailing reliance on poststructuralism, with its reification of discourse and avoidance of the structures and struggles of oppression obscures the origins, meanings and consequences of historical events. He observes:

> Those who opt for poststructural thought and privilege discourse bear no special blame for this academic drift. But they are part of the general process. Inasmuch as their writing and its conceptual ordering make few concessions to the needs and character of a reading public, indeed often revel in its aestheticized bombast, the claims made for a politics informed by language usually remain isolated within a narrow academic discourse itself . . .

(1990: 204)

This is more than simply a stylistic critique: I am concerned by the difficulty of what Palmer calls "aestheticized bombast" on women (men and children) who struggle with harsh economic realities, political uncertainties and social disruptions. How is it relevant for a Third World reading public, as well as development policy-makers, activists and other actors who are engaged with "development"? The paradox of textual analysis is particularly poignant where literacy remains a function of privilege. While it may secure women from the Third World (such as myself) privileged positions within academe, its political implications for poor women (and men) need to be explored.

Moving on to more substantive concerns, it is necessary to review Mohanty's and Ong's theoretical contributions to the debate on women in the Third World. Their deconstruction of the humanist project implicit in both liberal and socialist feminist scholarship is accomplished through a critical examination of three distinctive analytic strategies in Western feminist discourse. These are essentialist constructions of the category Woman, universalist assumptions of sexist oppression across cultures and, finally, political implications of essentialist and universalist suppositions that ultimately point to the colonialist intentions of Western feminist discourses.

They argue that empirical work on women in the South has not transcended Orientalism's preoccupation with essentialism. Western humanist lenses have interpreted Women in the Third World as powerless, passive, or to use the catch-all word, the "other." They believe the image of Third World women is essentialized through these representations and that these ahistorical constructs homogenize women's experiences based on the assumption of universally shared sexist oppression. Non-Western women constitute fixed essentialized categories expressed in their sexual or natural capacities. In particular, Mohanty argues, this analytic strategy is used where women are seen as victims of male violence, the colonial process, the Arab family process, the Islamic code or economic "development." As an undifferentiated category their experiences are shaped by a singular influence, i.e. the political economy. The WID approach in particular earns their ire for its focus on describing women's experiences and creating economic development policies in the light of universal Western assumptions about women's experiences. But they are principally concerned with whether socialist or capitalist versions of modernization emancipates or reinforces inequality. They point out that both of these approaches invoke binary constructions, which compare Western modernity with Third World traditionalism. As Ong elaborates:

> By using a traditional/modernity framework, these feminists view the destruction of "traditional customs" as either a decline of women's status in a romanticized "natural" economy, or as their liberation by Western economic rationality. This either/or argument reveals a kind of magical thinking about modernity which has proliferated in Third World governments, while confusing and obscuring the social meanings of change for people caught up in it.
>
> (1988: 83)

Within such a binary analytic, then, women can only be emancipated through Western economic rationality. Women are merely passive victims in this process. Mohanty argues, and rightly so, that such objectifications need to be continually challenged (see chapter 1).

Second, various methodologies serve to illustrate the "universal cross cultural operation of male dominance and female exploitation" (Mohanty 1991a: 66). Simple arithmetic can be used. For example the greater the number of women wearing the veil, the more one can claim universal sexual segregation and control over women. Second, a lack of historical and cultural specificity regarding concepts such as reproduction, the sexual division of labor, family, marriage household and patriarchy occlude differences in the experiences of men and women and obscure contradictory and complex social meanings attached to work (Ong 1988).

These methodologies, according to Mohanty and Ong, construct universalist generalizations instead of a historically informed, culturally specific analysis, and thus continue to colonize women in the South. Fran Hosken, for example, equates *purdah* in Muslim societies with rape, forced prostitution, polygamy, genital mutilation, pornography, and beating of girls (1981). She thus asserts the sexual control function of *purdah* and denies historical and cultural specificity to this mode of dressing. Mohanty argues that "It is in the analytic leap from the practice of veiling to an assertion of its general significance in controlling women that must be questioned" (Mohanty 1991a: 66). She points out the transformation in the veil's meaning – during the 1979 revolution Iranian middle-class women veiled themselves to indicate solidarity with their working-class sisters, while in present-day Iran, mandatory laws compel women to veil themselves.

Ong (1988) develops the same theme in her critique of Nash and Fernandez-Kelly's *Women, Men, and the International Division of Labor* (1983). The emphasis, she argues, is on men and women as a cheap source of labor, not on the meanings attached to work on the assembly line in multinational factories. In other words, the organizing theoretical principle is political economy rather than the cultural context. While Ong's point is important, it is also important to remember that this book was one of the first studies to reveal the impact of global restructuring on the international division of labor. It also dispelled the "nimble fingers" myth associated with Third World women engaged in labor-intensive assembly-line production, and proved the political and economic imperatives of this phenomenon. To reject such an analysis on hindsight, because it is now viewed with a newly obtained theoretical perspective, is to throw the baby out with the bathwater.

A more general consequence of universalism is the assumption that gender constitutes a superordinate category which can organize social relations across cultures. The result is a relentless determination by Western feminist discourses to find universal subordination of women in Third World societies, and to blur distinctions among women.

Finally, Mohanty shows that writings on women in the South provide a powerful political means to perpetuate colonialist relations between the North and South. The binary representation of First World/Third World women establishes a world order where the Third World women have failed to reach the evolutionary pinnacle of the First World. Mohanty's arguments clarify the colonialist motive in such writing:

> It limits theoretical analysis as well as reinforces Western cultural imperialism. For in the context of a first/third world balance of power, feminist analyses which perpetrate and sustain the hegemony of the idea of the superiority of the West produce a corresponding set of universal images of the "third world woman," images such as the veiled woman, the powerful mother, the chaste virgin, the obedient wife, etc. These images exist in universal, ahistorical splendor, setting in motion a colonialist discourse which exercises a very specific

164

power in defining, coding, and maintaining existing first/third world connections.

(1991a: 73)

ESSENTIALISM, DEVELOPMENT POLICY AND PRACTICE

Contemporary analyses of women in the Third World, at least in the post-1970 period, have been firmly rooted in policy-oriented research. Ester Boserup's study *Women's Role in Economic Development* (1970) and the inauguration of the UN Decade for Women were significant milestones for this body of scholarship. As Bandarage (1984: 497) notes, within just over a decade, the followers of the WID school produced a substantial body of knowledge on women's economic contributions and conditions in the Third World. Although widely critiqued for its liberal perspective, WID has still sought political change for women. Even socialist feminists/Marxist feminists have been concerned with the links between theory and practical action. Bandarage, who attempts to confront Marxist feminism's theoretical impasse, recognizes that "the combination of long-term structural change strategies with short-term solutions for meeting basic human needs, requires greater and more serious attention from Marxists" (1984: 505).

Given this framework, the following section raises some questions about Mohanty's and Ong's analyses. I reflect on three concerns. First, I discuss feminist emancipatory projects and their relationship to postmodernism. This discussion frames my second concern about the discourses on women in the Third World and their relationship to development policy and practice, particularly the improvement of women's lives. Finally, I discuss Mohanty, and Ong's noteworthy silence about the political nature of counter-discourses and resistances which preceded postmodernism, arguing that this silence, in effect manages to convey postmodernism's pre-eminent position in politically-conscious scholarship.

TEXTUAL ANALYSIS AND CONTEMPORARY ACADEMIC FEMINISM

In pursuing the question of the relevance of academic feminism to social change, postmodernist critiques pose certain dilemmas. My critique of these writings stems from two concerns. First, are we to believe that only postmodernism can lead the way to understanding the politics of "difference," based on local, contextual experiences? Have other forms of social theory and criticism been particularly ineffective in understanding experiences of women in the Third World? Second, can

165

postmodernist analysis lead the way to political action and change through a textual analysis that is decipherable for the most part only by erudite academic feminists? Does an overemphasis on language and discourse make the relationship to feminist emancipatory projects complex and difficult to comprehend (see also chapter 7)?

Textual discourse lays the groundwork for Ong's and Mohanty's theorizing about contemporary writing on women in the South. They argue that colonialist politics is a forceful presence in various texts written on women in the South. They focus on the political move present in Western feminist texts through which essentialist and univers- alist constructions of the Third World woman in Western feminist discourse achieves two objectives – first it limits political potential, and, second, it provides the colonialist move by confirming that women in the Third World can never rise above their subordinate status. Binary constructions of male and female inhibit analysis of power relations because power is always conceived as a male/female dyad. Women can only exert power if men lose it (Mohanty 1991a: 71).

Mohanty and Ong are particularly concerned with the colonialist implications of such discourses. The imperialist move is made when powerless women of the Third World are situated in contemporary Western feminist scholarship, unlike Western women who are/become real agents in postpatriarchal history. I share their concerns with the essentialism and universalism in the discourse on women in the Third World, and the political implications of these representations. But the concern with essentialism and universalism overemphasizes the impor- tance of textual analysis. As Gewertz and Errington (1991) argue:

we are worried that the *textual focus* . . . has the political implication of rendering virtually irrelevant to us the lives that actual – non- generic – "others" in fact lead . . . that our renderings of them determine their existence for us establishes them as products of our imaginations. This places them, in effect, as inhabiting realms essentially disconnected from our own. (We live in the world; they live in our imagination) . . . It may curtail our understanding of those socio-historical forces of systemic connection, those forces which articulate between and shape our lives and theirs in a world system.

(1991: 81)

The disjunction between "their" lives and "our" lives created by textual analysis renders political projects impractical. If there is no connectedness between the two realms, "us" and "other," then how is it possible to form strategic coalitions across class, race, and national boundaries? I believe theoretical strategies that show the interconnect-

166

edness of lives in a world-system are needed before coalitions can be envisioned and attempted (see chapter 7).

Such textual analysis of women in the South compels a re-examination of feminist theory and projects because the WID school and its socialist feminist counterpart have been linked since their inception with praxis. We need to unravel whether/how postmodernist feminist analyses can lead to practical action since an overriding concern with postmodern analyses has been its potential (or lack of) for political action (Hartsock 1989; Nicholson 1990). The question is especially significant for feminist studies as its central project pertains to gender politics. A common concern is that postmodernism cannot provide a theory of change because it regards all (existing) knowledge as invalid and debased by Western humanist projects. The deep commitment to political action which is evident in both Mohanty's and Ong's writings, seems an inexplicable contradiction to reject the philosophical underpinnings that sustain and enable their feminist projects. For, as Rita Felski argues, the political nature of feminist studies is itself derived from the Enlightenment:

> Drawing on an egalitarian social imaginary which is a product of the democratic aspirations of the Enlightenment, feminists have also radicalized this tradition by turning it against itself, demonstrating how its attempted actualization has in fact engendered new and deeply ingrained forms of gender subordination. Rather than announcing the death of rationality, subjectivity or history, feminist practices indicate that such concepts must be thought differently in relation to the struggles of gender politics.

(1989: 53)

But Ong and Mohanty reject social criticism that depends upon Western-oriented philosophical underpinnings. This reflects postmodernists' assertion that philosophy can no longer claim to ground social criticism: criticism should be local, *ad hoc* and untheoretical. Large-scale historical narrative and social theoretical analyses of domination and subordination thus become illegitimate genres. Since feminism's ultimate concern is with social change, postmodernism must keep open the possibility that some forms of knowledge are emancipatory and are likely to be appraised for their potential for political change. Unless they are able to establish a political potential, and in the case of women in the South, confront some harsh realities that force attention to policy issues Mohanty and Ong can be accused of falling into the trap of academic feminism, which, as Currie and Kazi (1987) note, has become separated from feminism as a social movement and lost its potential for social change.

Within the context of such a development, postmodern feminist

167

analyses of women in the Third World raise some important questions. Has the postmodernist version of feminist theory, to voice Mary Evans' (1982) concern, been transformed into an academic debate which no longer has relevance for women outside the hallowed halls of academic institutions? Has women's studies become the new road to elitism, writing and thinking for the privileged few (Bowles and Duelli-Klein 1983)? While these concerns were voiced before postmodernism took hold, current concerns about postmodernism's privileged positioning in academia abound (Harvey 1989; O'Hanlon and Washbrook 1992; Palmer 1990).

Most importantly, does postmodernism require feminism to abandon the category of women and the proposition that they have their own history? If so, it dissolves its own subject. As Christine Di Stefano (1990a) points out, although the postmodernist critique against universal categories of women is well taken, feminists cannot entirely reject such categories since the world for the most part continues to treat women on this basis (Denise Riley 1988: 112–14). The way out of this theoretical impasse is provided by Fraser and Nicholson:

> A first step is to recognize, . . . that postmodern critique need forswear neither large historical narratives nor analyses of societal macrostructures. This point is important for feminists, since sexism has a long history and is deeply and pervasively embedded in contemporary societies. Thus, postmodern feminists need not abandon the large theoretical tools needed to address large political problems.
>
> (1990: 34)

Indeed, Scott and Butler (1992) reject the notion that politics requires universal values. They, along with a number of other scholars (Canning 1994; Fraser and Nicholson 1990; Sylvester 1994) believe postmodern feminist critiques will have to be infused with temporality and historically and culturally specific institutional categories (see Introduction/ Conclusion) if they are to have political relevance. It is precisely the combination of essentialism, as it gets explicated in meta-theory, and historically contingent analysis, that is taken up for discussion in the next section.

ESSENTIALISM, META-THEORY AND POLITICS

The emphasis on textual analysis leads to certain theoretical dead ends as deconstruction and delegitimizing essentialist and universalist forms of argument do not point to any basis for reasoned action. As David Harvey argues, denying . . . metatheory which can grasp the political–economic processes (money flows, international divisions of labor, financial markets, and the like) that are becoming even more universalizing in their depth, intensity, reach and power over daily life. In its rejection of totalizing it [postmodernism] avoids the realities of a global political economy and the nature of global power.

(1989: 117)

Mohanty in particular falls into this trap. While she rejects totalizing textual practices, she remains committed to feminist political goals. Yet, postmodernist analysis by itself offers little by way of potential for political action, especially if the experiences of women in the South are understood as local, fragmented and contextual. How can we build coalitions unless we construct some universalizing experiences that bond us together? Mohanty's political interest in coalition-building also implies that we should pay attention to concerns that flow from racial/gender/class subordination, all of which can be viewed as unitary, essentialized categories.

How are we able to do that? In practical terms, we need international coalitions to unify our fragmented experiences and to provide a sense of shared experiences. But this means lobbying for humane policies, and attempts to influence public policy, both at the domestic and international level. How can we accommodate Mohanty's political projects without resorting to universalism for the sake of public policy? The immediate practical concern of defining a similar form of injustice as a unitary notion shared by disparate groups of people requires that we do not reject essentialist understandings completely when we move into public policy (see chapter 7).

In concentrating on textual discourse, rejection of essentialism and universalism provides an escape from the moral dilemma intrinsic to social criticism and social theory, particularly in regard to transformation in the South. What makes much postmodernism politically bankrupt outside the academic context is its failure to confront the moral underpinnings of development policy and practice such as literacy programs and expansion of basic healthcare facilities.

The preoccupation with anti-essentialism makes this moral dilemma a central concern in understanding the experiences of women in the Third World. As Nussbaum argues:

Essentialism is becoming a dirty word in the academy and in those parts of human life that are influenced by it. Essentialism – which for these purposes I shall understand as the view that human life has certain defining features – is linked by its opponents with an ignorance of history, with lack of sensitivity to the voices of women and minorities. It is taken usually without extended argument, to be

169

in league with racism and sexism, with "patriarchal" thinking
generally, where extreme relativism is taken to be a recipe for social
progress.

(1992: 205)

Nussbaum argues that essentialism is rejected without any philosophical conceptualization of what is good or bad for women. For anti-
essentialism can include accepting traditions such as ancient religious
taboos, the luxury of the pampered husband, ill health, ignorance and
death. Nussbaum clarifies this in her account of the experiences of
women development workers in literacy projects in Bangladesh (see
Chen 1983). She shows how essentialism and local contextual understandings can be joined to bring about significant and positive changes in
the lives of women.

> . . . none of these concrete transformations could have happened
> had the women of the development agency not held fast to their
> general conception, showing the women its many other concrete
> realizations and proceeding with confidence that it did have some
> concrete realization in these particular lives. Essentialism and parti-
> cular perception were not opposed: they were complementary
> aspects of a single process of deliberation. Had the women not been
> seen as a human beings who shared with other women a common
> humanity, the local women could not have told their story they did,
> nor could development workers have brought their own experiences
> of feminism into participatory dialogue as if they had some relevance
> for the local women. The very structure of the dialogue presupposed
> the recognition of common humanity, and it was only with this basis
> securely established that they could fruitfully explore the concrete
> circumstances in which they were trying, in the one case, to live and
> in the other case, to promote flourishing lives.

(1992: 236–7)

Nussbaum's insights into anti-essentialism and development practice
are particularly illuminating: she is able to unpack the moral dimension
that undergirds theorizing on the Third World. An enduring concern of
modernization and underdevelopment theory has been an improvement
in the quality of life. It is with the sensibility obtained from Nussbaum's
analysis that we learn that anti-essentialist arguments need to be constrained by political pragmatism (see chapter 10). For example, Mohanty
ignores the benefits of joining essentialism and historically specific
analysis in her appreciative assessment of Mies's work on lace-makers
in Narsapur, India. She overemphasizes the historically and culturally
specific analysis of Mies for the convenience of her theoretical argument,
while ignoring the fact that Mies's study is situated within a totalizing

concept of the world market. The title of Mies's work *The Lace Makers of Narsapur: Indian Housewives Produce for the World Market* stands testimony to that. Indeed, the meta-theory of world-systems analysis frames her book. Just as the analysis of the literacy program in Bangladesh demonstrated, Mies shows that an essentialized concept of the world market can be combined with a local contextual analysis. Interestingly, Mohanty, who is concerned with the politics of representation and the Western woman's colonialist move in constructing the "other," ignores certain telling remarks of Mies's location in this politically sensitive analysis. Sounding remarkably like a middle-class woman complaining about the lack of "good help," Maria Mies begins her acknowledgments:

> Everyone who has done empirical research on women in rural India knows it is not easy to find women investigators who are ready to go to the villages and share the rough life of peasant women. In spite of educational and academic qualifications many urban women are still so handicapped by patriarchal norms and institutions that they dare not move beyond the radius of cities and towns. I was very fortunate, therefore, to find two young women, Lalita and Krishna Kumari who were courageous enough to come along to Narsapur, to live in the villages among the lace-making women and share their life.

Interestingly, Lalita and Krishna Kumari disappear from the authorial signatures of the text – it is Maria Mies alone who writes about Indian housewives – again an essentialist category, for are they not Agnikulak-shatriya, Christian and Kapu women who are primarily engaged in lace-making?

COUNTER-DISCOURSES, ACTIVISM AND POLITICS

A notable feature of postmodern analyses of development[6] is the curious silence about the political changes wrought by earlier resistance to modernization, in counter discourses such as dependency theory, world systems theory or women in development/socialist feminism, both within and outside academia. This silence implies that only postmodernism is a legitimate genre of social critique. Outside the academic institution, social movements and resistances against modernization or colonialist projects have provided a powerful attack on essentializing discourses about the South. By ignoring the transformative effect of popular protest, Mohanty and Ong manage to portray postmodernist analyses as the only powerful challenge to essentialism. Mohanty's reference to previous work on women in the Third World – "I do not question the descriptive and informative value of most Western feminist writings on women in the third world" – is only in passing and somewhat inadequate (1991a: 54).

171

Rejection of all other discourses on development raises two issues or questions. First, does this mean that critiques, such as dependency theory and world systems theory do/have not accomplished their political goals? Are both liberal and socialist feminist versions of Gender and Development political failures? This is difficult to argue as WID gave voice to women and dragged the issues of gender onto center stage in the international aid regime. It also provided a forum for women in the three world conferences. More importantly, the WID paradigm created political space for women in academia to pursue feminist work on the Third World as legitimate academic activity. WID and GAD have fueled a virtual explosion of work done on women in the South, undergraduate and graduate courses taught in universities, and dissertations written in this new subdiscipline.[7] Second, has the relationship of policy to the scholarship on women in the South been a complete failure? Not so judging from the plethora of studies that examine the relationship of social and economic transformation and gender relations. However problematic the essentialized conception of women, the constitution of gender as a category of analysis even in mainstream academic scholarship is a political achievement. It has shifted analysis from an androcentric stance to a greater focus on women's experiences.

While Mohanty and Ong caution against constituting gender as a superordinate category of analysis, this sensibility is also shared by social scientists working outside the postmodernist frameworks who acknowledge that it need not be so:

> while we have no argument with the fact that gender is a potentially valid dimension along which to examine the equity and efficiency effects of economic reform and adjustment, we would like to see the utility of this approach scrutinized more closely. When is gender the most appropriate first-order disaggregation? Under what circumstances would we better understand the fundamental, structural cause of the problem by examining distributional impacts across all landholdings, region, tenure status, employment status, or occupational sector? When, for any of these first-order disaggregations is gender a second order disaggregation? Gender should not be an automatic starting point of analysis of adjustment [structural], nor should it be ignored as a potential starting point. We encourage the avoidance of approaches that are gender-blinded or gender-blind. In general, we caution against extreme views.
>
> (Sahn and Haddad 1991: 1448)

This passage is informative on several counts. First, it tells us that gender as a category of analysis is of general concern, not just for feminists, but also for social scientists working with policy-oriented issues. Second, Sahn and Haddad remind us that gender does not constitute a super-ordinate category, but takes precedence only within certain situations. At other times, it may be necessary to take into account variables that become the organizing principle for a particular experience. Finally, they arrive at the same conclusion as Mohanty and Ong about the dangers of universalist and essentialist categories, yet this is not derived from a textual analysis, but from inductive methodological strategies based on the varied experiences of women. They proceed to use their insights as a beginning premise for their analysis of women's experiences with structural adjustment programs.

In privileging postmodernism as social criticism par excellence, Ong and Mohanty do a disservice to feminist scholarship which has brought sweeping changes to social theory. To use Mohanty's own description of feminist scholarship, "it can best be seen as a mode of intervention into particular hegemonic discourses, . . . it is a political praxis which counters and resists the totalizing imperative of age old 'legitimate' and scientific bodies of knowledge" (1991a: 54). Mohanty and Ong imply that their genre of feminist scholarship is the best form of social criticism. They accuse WID and socialist feminism of reinforcing colonialist/neo-colonialist relationships and failing as a political practice. This is shortsighted. To steal a page from the postmodernist's book, perhaps we need multiple genres of scholarship to enrich social criticism and to make it meaningful in different contexts.

Moreover, the focus on essentialism and universalizing of women's experiences in the Third World ignores the number of resistances organized by women in various locations in the South. Third World women have not merely been passive actors, immobilized by Western representations. Women in the South had been mobilizing protests on a large scale long before the UN declared a Decade for Women. One of the most vocal critics of environmental modernization who has challenged destructive ecological practices, Vandana Shiva (who herself is a challenge to essentializing conceptions of the "object" Third World woman status) describes the Chipko movement, an ecological movement to protect the forests of the Himalayas that emerged more than three hundred years ago, and which continues to this day:

> The Chipko process as a resurgence of woman power and ecological concern in the Garhwal Himalayas is a similar mosaic of many events and multiple actors. The significant catalysers of the transformations which made Chipko resistance possible have been woman like Mira Behn, Sarla Behn, Bimala Behn, Hima Devi, Gauri Devi, Gunga Devi, Bachni Devi, Itwari Devi, Chamun Devi and many others. The men of the movement like Sunderlal Bahuguna, Chandi

rasad Bhatt, Ghanshyam Shailani and Dhoom Singh, Negi have
en their students and followers.

(1988: 68)

Sunderlal Bahuguna, the most well known male leader associated with
the movement, acknowledges the women's primary role: "We are the
runners and messengers – the real leaders are the woman" (cited in
Shiva 1988: 70).

In Trinidad and Tobago, where free trade zones (FTZs) have been
established as an ideal type of unregulated trade, women have resisted
the working conditions in the industries and questioned the value of
these economic policies to development. McAfee, in detailing the experiences of women under structural adjustment programs in the Caribbean
region cites Asha Kambon of Women Against Free Trade Zones:

> We do have to export, but the questions are: what do we export?
> Why, where, to whom, on what terms? We do need hard work, and
> we need to make sacrifices, but who is sacrificing, and what do we
> gain? In Trinidad and Tobago, we have challenged the policy of
> export processing zones not only because of their exploitation of
> women, but also [because of] their questionable role in national
> development. Now we take up every issue that comes to national
> attention in terms of whether or not and how it contributes to or
> detracts from development.
>
> (cited in McAfee 1991: 88)

McAfee's analysis, which is framed within a Marxist political economy
approach, problematizes the changes brought about by unequal relations in the world economy. But within this totalizing narrative she has
been able to show women's resistance to subjugating economic and
political practices. This has involved collaboration with women in other
Caribbean nations. As McAfee explains:

> The struggle of Jamaican women against exploitation in the FTZs
> continues. As word of their movement has spread, they have linked
> their efforts with women in other Caribbean countries who face
> similar conditions. . . . The plan of the government of Trinidad
> and Tobago, pressed by USAID [United States Agency for Interna-
> tional Development] and the multilateral lenders to establish EPZs
> [Export Processing Zones] in Trinidad, has met with strong protests
> by Trinidadian women. Because of their ties to the Caribbean
> regional women's movement, women in Trinidad are already
> familiar with the consequence of such a strategy.
>
> (1991: 88–9)

174

The coalitions between women of different nations within the Caribbean region were built through a belief in the global nature of imperialist forces and an understanding of their shared oppression and subject status as women. Mobilization without this meta-understanding is hard to imagine. Postmodernism has been criticized for its reliance on relativism and its rejection of grand theory. As Nicholson poses the question: Are coherent theory and politics possible within a postmodern position? (1990: 9). It is possible, Nicholson responds, if postmodernism does not eliminate all grand theory, but undertakes theorizing which is explicitly historical, that is, located within historically specific contexts. Here Butler and Scott's (1992) call for political action based on carefully situated, partial knowledges may provide an opening for a postmodern feminist political action to avoid false generalizations. However Mohanty and Ong do not adopt this position. Their analysis of writing on women in the Third World overemphasizes the problems associated with essentialism and universalism, while downplaying the problems their approach creates for dealing with the political aspects and lived reality of the women confronting the dehumanizing experiences of modernization.

CONCLUSION

The above critique is not meant to be an exhaustive review of recent postmodernist analysis on the Third World. That would be a larger project, beyond the scope of this chapter. Neither do I claim to give any definitive answers to feminist ethnography's question: whose voice? whose representations? or as Patai (1991) poses "US academics and Third World women: Is ethical research possible?" Instead, I am reacting to a particular aspect of contemporary postmodernist analyses of Third World women, particularly to Mohanty's and Ong's stand against the political impact of previous scholarship, which in effect positions their perspective as social criticism *par excellence.* They imply that only they can confront "difference," since other genres of social criticism fail to move past essentialism and universalism. But in putting forth these claims they successfully evade the moral issues of poverty, hunger, inadequate health care and lack of literacy which have historically been of central concern to the scholarship on Women and Development in the South.

These concerns frame my critique of postmodernism, particularly the need to be cautious in pursuing textual analysis as our primary theoretical strategy. As I have shown, social criticism fails to make the connection to praxis through a textual analysis. The emphasis on essentialism and universalism in creating the Us/Other categories creates a disjunction that seems impossible to bridge. It is this that makes postmodernism's link to practical action problematic. In order to break through this impasse, I believe we cannot entirely eliminate essentialist categories, especially as feminism is founded on a category "woman." While acknowledging differences among women, we must devise theorizing strategies to unite our diverse experiences for political coalitions and alliances. A combination of broad generalizations (essentialism) carefully constructed with a historically contingent analysis is needed if we are to confront some of the moral issues that drives much of development policy and practice in the Third World. Indeed, current writings by postmodern feminists such as Butler and Scott (1992), Canning (1994) and Sylvester (1994), suggest a move in that direction.

A degree of humility would be beneficial as well. Postmodernists are not the first to acknowledge "difference." Activists and policy-oriented researchers have often discussed differences, but from "on-the-ground experiences." Although within academic scholarship postmodernism has made us attentive to "difference," we should acknowledge the contribution of descriptive accounts that succeed in doing the same. In particular, women's movements and resistances constitute a powerful strike against essentialist or colonialist understandings of Western feminist writing. This is now passing into common sensibility, not shared exclusively by those working in the postmodern milieu. I suggest we remember that, and seek theoretical flexibility while keeping the moral and practical concerns of our emancipatory projects well within sight.

NOTES

1 Earlier versions of Mohanty's essay appeared in *Boundary* 2(12)3/1, (Spring/ Fall 1984) and *Feminist Review* 30 (Autumn 1988).
2 Modernization and development are representative of the school of thought that perceives the expansion of capitalism as a means to progress. The term underdevelopment includes the linked but different perspectives of classical imperialist, dependency, urban-bias and world systems accounts that critiques the expansion of capitalism as a blueprint for social progress.
3 Robert E. Wood's term in *From Marshall Plan to Debt Crisis* (1986).
4 Bandarage (1984) provides a fine overview of WID from a socialist feminist perspective.
5 See O'Hanlon and Washbrook (1992). They ask, "Why, then, have these perspectives achieved such widespread popularity in Western, particularly American, academic circles? There is now, of course, a large and influential body of postmodernist writing in history and anthropology, mostly published in the United States" (p. 158).
6 Here I include Arturo Escobar's (1984–5) analysis of discourse and power in development. One means of circumventing the hegemonic deployment of development is participatory action research. "From a philosophical point of view, participatory action research constitutes a radical departure from traditional Western philosophy. By its very nature, it rejects the subject/

object division, central to Western philosophy and empiricist social science, and transforms them into a subject project" (391–2).

7 At the University of California, Davis, for instance, there are at least four undergraduate courses taught by the Departments of Anthropology, Geography, History and Sociology on gender-related issues in the Third World, and at least three graduate seminars offered by the Departments of Anthropology, Geography, and Sociology. This does not include other feminist theory seminars and courses.

Part V

KNOWLEDGE, DEVELOPMENT AND PRAXIS

In this last part, Christine Sylvester, Eva Rathgeber and Jane Parpart address some of postmodern feminism's concerns (including the issues of representation, identity and knowledge raised by the other contributors) within the context of development practice. In looking more closely at the development enterprise itself, they go beyond the (theoretical) postmodern feminist critiques of development theory. In so doing, the authors show that postmodern feminist thought has implications for development practitioners as well.

As shown in previous chapters, the representation of Third World women as passive, ignorant, voiceless and tradition-bound objects (partially) constructs Western women as modern subjects in control of their environment and their futures. The dichotomy between "rational, emancipated Western woman" and "emotional, passive Third World woman" acquires an added dimension when we consider the development enterprise. With development theory and practice firmly embedded in Enlightenment thought, a space has been created for the development specialist. This Western (-educated) development expert is the gatekeeper to and the holder of specialized knowledge about the development enterprise. In other words, one could argue that the (developmental) knowledge/power nexus has found its embodiment in the development expert.

The prioritizing of Western knowledge and Western (-educated) development experts has various implications. It has, for instance, resulted in the dichotomization between modern, technocratic knowledge and traditional, indigenous expertise. Due to this dichotomization, the latter has often been silenced, ignored or subjugated. Quite frequently this has led to the designing of (mal-)development projects without a clear understanding of the local situation.

Since the 1970s, alternative development advocates have tried to change this situation with varied success. Unfortunately, many mainstream development practitioners and agencies have not really ques-

179

tioned the role of the development expert and, least of all, the power dimensions embedded in the development enterprise.

Sylvester, Rathgeber and Parpart focus on opposite dimensions of the development enterprise: local women's development expertise and the role of the gender and development expert. In her chapter, Sylvester argues that in (rural) Zimbabwe various power relations or regimes, ranging from a "Marxist promise" to one of "liberal pragmatism," intersect with a gender regime. The construction of Zimbabwean women varies according to which regimes intersect. And as Sylvester argues, because "regimes of truth are multiple, overlapping, and internally inconsistent, subjects can mediate power and knowledge at various margins." The Zimbabwean women were able to get access to "Western" knowledge (i.e. special courses, education) and for many this exposure buttressed their personal power. However, because the women did not value their own indigenous development expertise very highly, they did not challenge the "Western" knowledge's legitimacy. In other words, the hierarchical relationship between Western-based development expertise and local knowledge was never fundamentally challenged.

In her contribution, Eva Rathgeber discusses the issues raised in Sylvester's chapter on the basis of her insider knowledge about the development enterprise. She finds that the perceived dichotomization between Western and indigenous knowledge is not true in practice. Although she recognizes that in the past the development expert has had too much power (especially within the WID approach), she argues that the GAD approach is flexible enough to incorporate local knowledge and expertise and thus to lead to a more nuanced, interactive form of development. She thus raises the possibility of a conjunction between GAD and postmodern feminism that could have important implications for development theory and practice.

In contrast, Jane Parpart is more sceptical about GAD's flexibility to incorporate indigenous knowledge and fundamentally challenges the "institution" of the development expert. Her analysis, (partially) informed by postmodernist feminist thought, emphasizes the modernist underpinnings of the development enterprise in general and the development expert in particular. She argues that the development expert exercises a disciplining power which contributes to the control, production and reproduction of the South. Although Parpart does not reject the institution of the development expert entirely, she thinks that a (fundamental) questioning of the modernist and patriarchal nature of Western knowledge should become part of development practice.

All three authors believe the current dichotomization and hierarchization between indigenous, traditional knowledge and modernist, technocratic expertise must change. How to achieve that goal remains in dispute, but the recognition of difference, the focus on language and power and the interrogation of "claims to know" are clearly central. For that, postmodern feminist insights, in one shape or another, can play an important role.

10

"WOMEN" IN RURAL PRODUCER GROUPS AND THE DIVERSE POLITICS OF TRUTH IN ZIMBABWE

Christine Sylvester[1]

Michel Foucault argued that ideology is "a notion that cannot be used without circumspection" (Rabinow 1984: 60) and offered three reasons for this. First, ideology "always stands in virtual opposition to something else which is supposed to count for truth" (60), something less mystified and more objective. Second, it is secondary to "something which functions as its infrastructure, as its material, economic determinant, etc." (60) – often the state and its juridical representations. Third, it refers to a subject who takes ideology in and is repressed or oppressed by it, or who dreams, suggest Alessandro Fontana and Pasquale Pasquino, of "a quasi-transparent form of knowledge, free from all error and illusion" (cited in Rabinow, 59). All of this poses ideology as part of "a" negative and coherent metapower. Foucault's ideas, by contrast, point us in the direction of regimes of truth that are produced "only by virtue of multiple forms of constraint and induce regular effects of power" (72–3) – good and bad, positive and negative. Such regimes stipulate:

> the types of discourse which [they] accept and make function as true; the mechanism and instances which enable one to distinguish true and false statements; the means by which each is sanctioned; the techniques and procedures accorded value in the acquisition of truth; the status of those who are charged with saying what counts as true.
>
> (Rabinow 1984: 73)

Zimbabwe is characterized by contestation and consent to many series of power relations that can be called regimes. At the level of national politics, I have identified a few of what are undoubtedly many overlapping and yet distinguishable regimes of truth. They are the regimes of Marxist promise, of liberal pragmatism, and of authoritarian strength

(Sylvester 1991b; 1986). Singly and together they intersect with an historically variegated gender regime that interweaves elements of indigenous "tradition" with colonial messages about the proper place of African and white women in Southern Rhodesia, adding more recent efforts to define the status of "women" in the new Zimbabwe.[2] There are also intersections of all the aforementioned regimes with a regime of aid and development that shapes people, including the "women" among them, into eligibility for modernity via funding policies.

The gender regime offers the most direct route to knowledge about people called women that is meant to "function as true." Each of the other regimes offers indirect guidelines on how such people should comport themselves in order to be appropriately Marxist, pragmatic, strongly indispensable, worthy of aid and so on. Invariably, all the regimes discursively evoke "women" as female-bodied persons who have specific and fixed tasks related to childbearing and rearing, nurturance, and support for persons called men. Over and beyond these tasks, "women" have other duties, and each regime specifies what those duties are. Some people accede to these functioning truths, some resist, and some produce their own knowledge in ways that ensure that power is not just the preserve of institutions formally established to shape and garner "it." This chapter explores the ways people called rural women define and constitute the "rules" of "women" and "women's production" in peasant farming groups and agricultural producer cooperatives in Mashonaland East Province.

The farming groups grow out of an historical practice of *mushandira pamwe*, working together to plow fields. The Rhodesian state harnessed this practice to a project of scientific production in the "native reserves," and Michael Drinkwater (1991) contends that Zimbabwe's agricultural policies still sound the notes of colonial conviction that African farmers must embrace and apply science to group endeavors in order to be truly productive (the group orientation to production is not urged on commercial farmers who cultivate on large-scale, privately-owned estates). With current public policies reproducing a colonial bias toward Western expertise that persists in subordinating the lifeworld of the peasant, it is not surprising, Drinkwater says, that rural Zimbabweans still resist state agrarian policy, even though the state is now African. The unaddressed question in his work is whether and how "women" figure into that resistance. How do they conceptualize gender and farm work? Which regimes of truth do they follow and/or reject in legitimating their relationships to farming? Are they constructing a new regime of truth about agricultural work relations, abiding by inherited rules of tradition or science, or operating somewhere in-between the old ways and a new horizon of knowledge?

Producer cooperatives are new to peasant areas. During the colonial era marketing and supply cooperatives serviced the commercial farming sector and one mixed-race producer cooperative existed at Cold Comfort Farm. It was only with independence and government encouragement that rural and urban cooperatives emerged. But officialdom quickly turned ambivalent. The (then) Prime Minister's office first took responsibility for cooperatives and then shifted the purview to the Ministry of Lands, Agriculture and Rural Resettlement. A Ministry of Cooperatives was formed in 1986 and was merged with the Ministry of Community Development and Women's Affairs in 1988, at which point "cooperatives" became identified with "women." By 1989 the government had hived off Cooperatives and Community Development from Women's Affairs and placed the latter under the Senior Minister of Political Affairs (a "man"). After rearranging government in 1992, cooperatives were assigned to a ministry of National Affairs and Employment Creation. The regular reshuffling of the cooperatives portfolio over the last ten years contrasts with the more stable approach to farming groups that the government has followed from colonial times. It may influence the ways "women cooperators" view themselves and their chosen venues of production relative to the more steadily aided farm groups.

GLIMPSING THE FRACTURED FIELD OF POLITICS IN ZIMBABWE

If gender is a regime-shaped construction, then it is sensible to view Zimbabwean women as "women," that is, as bearers of an unsettled, unfixed and indeterminate subject status that the people thus labeled may or may not embrace. This beginning point means that we cannot assume *a priori* that people who wear skirts, bear and suckle children, carry water and so forth, are genuine women. We can only assume that gender is an historically contingent set of local social assignments that we must discern and query, just as many of us routinely query terms like "development" or "progress." It seems obvious to say that people are not necessarily what they are called. Yet to query "women" can sound unnecessary and strange because women is so often a given.[3] Surely all of us know women when we see them! But do we? Whose notions of "women" guide our vision and potentially freeze people in relation to their bodies and usual social assignments? What do people called women call themselves? Are there gaps between the usual confident understanding of women we speak about and the self-understandings that people negotiate for themselves in local contexts? Are Zimbabwean "women" in farming groups the same people who are "women" in cooperatives? Should we not problematize this crucial category of meaning and identity instead of accepting it as truly pre-given, especially in settings

184

where there are competing messages about how one should properly inhabit gender?[4]

Each regime of truth in Zimbabwe overlaps and unfolds in, on, and around the others in ways that render it less than fully coherent. Yet in the amalgamation of space, time and meaning that produce a common sense of women, there are webs of opinion that interlock, as the following discussion indicates.

THE REGIME OF PROMISE

The regime of promise produces "multiple and indefinite power relations" that address the fading but not eviscerated dream of a producer-run society overseen by a Marxist–Leninist ruling party (ZANU-PF). Centralized in the person of Robert Mugabe, a few party cronies, some members of collective cooperatives, and lingering leftists, the regime of promise was "the" discursive promotion of the first eight years of Zimbabwe's national existence. The promises, however, did not turn into consistent public policies (Sylvester 1990a; 1986). If they had, some rural cultivators might have gained status as "workers" in a peasant "class," while those forming producer cooperatives would have become "socialist man." Presumably "women" would have evacuated private spaces to become an identity associated with a life of production outside the home. Yet the regime of promise often undercut its own producer message by writing of "women" as "mothers" and "family"-bound creatures first and foremost, saying that women have the decisive role "in maintaining the *family* as the basic social unit of our society" (ZANU-PF 1990: 21, emphasis in original), and that "the extended family derives its strength and vitality from the women" (21). These statements gave "men" tacit permission to be in other places and to leave "women" weighed down as they tried to enhance their promise-scripted capacity to be "workers" and "producers" in farming groups or cooperatives. Although this regime has been eclipsed by others, its rendering of "women" commonly reappears in other politics of truth.

THE REGIME OF STRENGTH

The regime of strength Zimbabweanizes the divisive power relations that marked the Rhodesian era. It makes "true" the notion that some people are indispensable to the well-being of the country and others are more easily ignored or even demonized. Thus, the government threatens at every turn to deprive white farmers of vast tracks of underutilized lands, but avoids doing anything precipitous to endanger a group presented as indispensable to the economy.[5] Indeed, "[p]lans are afoot to promote black large-scale commercial farmers to rival some 4,000 white farmers responsible for at least 82 per cent of the Z$2.2 billion total revenue from Zimbabwe's agricultural output" (Sakaike 1991: 10). Meanwhile, peasants wait and wait to recover lands appropriated by the Rhodesians, sometimes dodging government retribution against "squatters" (Palmer 1990).

The regime of strength is associated with a black and white elite that promotes hierarchical relations of place. Commercial farmers and government officials – who tend also to be "men" – have elevated place. The place of most "women," black and white, is as loyal, obedient and decorous ones. When they step outside the rules of what some feminists would surely detect as a patriarchal set of power relations embedded in the regime of strength, "women" can find themselves condemned as "libbers" or, if black, rounded up as "prostitutes."[6] The overriding truth of this regime is that everyone must appear contentedly in an externally-defined place, without any show of "dissidence."

And so it is for rural producer groups too. According to purveyors of government expertise, cooperatives must conform to rigid, outer-directed rules. They must be "association[s] of free and equal individuals for social and economic gain," wherein "everyone receives equal pay for equal work and all decisions are made by all members of the cooperative" (Mathema and Staunton 1985: 1). They must uphold six general principles of cooperative organization and organize a hierarchical management structure. That "males' representation [on cooperative committees] [wa]s approximately 20 per cent higher than their representation within total membership numbers" in 1987 (Chinemana 1987: 15) would be of no concern to representatives of strength; nor would complaints by "women members" that they work "harder than men since at home [we] do domestic work and men do nothing" (20). In this regime, people must know their place and accept it.

Farming groups are also rigidly structured under a Group Development Area scheme that is overseen by the agricultural extension service of Zimbabwe (Agritex). Several communal area villages form Agritex-affiliated committees and send representatives to an umbrella committee for the area, a ward committee, and a district agricultural committee. The local units of the scheme are the farming groups that bring together:

> six to eight people who live near each other and help each other plow, harvest, etc. They plan an itinerary to do one person's plowing on Monday, another person's on Tuesday and so on. They are also expected to organize their own transport and fertilizers and work together. The successful ones wear uniforms, organize singing choirs, compose their own songs with messages to promote, and sometimes compete.

> (official from Agritex)

Lending is also an area where the regime of strength flexes its muscles:

> Our application process is quite exhaustive. We calculate yield levels
> and ability to repay on the basis of past performance. . . . the biggest
> problem is people avoiding paying the AFC [Agricultural Finance
> Corporation] by selling their produce to neighbors or selling it
> locally where we have no ability to put a stop-order on their
> profits, as we do when they sell through the marketing boards . . .
> We have to have Agritex tell us how much farmers should have
> gotten in yield and whether their excuse is realistic. In 1984–5 we
> refused most second loans.
>
> (AFC official)

Funds are available for everything from fencing and farm implements
to draft animals. But "if it's a particularly good client, we even have
money for school fees. This is usually the case for commercial farmers.
. . . We don't do this for everyone because we don't want to burden
people with too much credit." It seems "the people" rely on groups and
government rules to get ahead and certain "indispensables" get loans.

THE REGIME OF PRAGMATISM

A third regime resonates with Southern Rhodesia's brief effort after
World War II to try on the politics of liberalism in order to balance
competing economic demands presented by variously situated whites
(e.g. those in manufacturing versus those in agriculture; workers versus
management) and by Africans seeking greater political and economic
power (see Sylvester 1991b, chapter 2). Today's regime similarly tolerates
political, economic and social disunities in Zimbabwe if they can be
balanced pragmatically through pluralist politics. A representative of a
donor organization in Harare told me that "[i]n Zimbabwe ideology
hasn't been an issue because government people are pragmatic, positive
and mature." Mugabe reinforces this view when he refers to "our
socialist practice over the last 11 years as pragmatic socialism" (quoted
in Sakaike 1991: 9), that is, a socialism that is no longer scientifically
Marxist and is, rather, more akin to the Fabian or democratic politics
that mark many contemporary venues of liberalism.

Pragmatism, the seeming preference of many average Zimbabweans,[7]
neither encourages nor discourages the existence of self-negotiated
"women." Officially this politics is "gender blind," in much the way
that it is blind to class problems in an increasingly corrupt Zimbabwe
and blind to its own earlier credo of growth with equity, now all but
abandoned in a "pragmatic" swing to structural adjustment. Indeed, a
regime of pragmatism can rest content with the appearance of neutrality

that comes from establishing *de jure* rules of and opportunities for equality, countenancing many *de facto* inequalities:

> In 1980 the government decided on equal opportunity for women and men and increased the number of places at agricultural colleges through quotas, etc. Some lady students came, but not in large numbers because of the tendency to educate the male children and there being not enough females with the required O Levels to get in.
>
> (Agritex official)

> All this emphasis on women . . . seems too much. I don't approve of separate activities for women. They can always just join the men.
>   Me: Can men always just join the women?
>   (No response).
> (Official of Mashonaland East Province, Ministry of Community and
>   Cooperative Development and Women's Affairs)

Under the power relations of pragmatism, the "truth" is that "modifications can only take place in a gradual way" (Zimbabwe, Government of 1982: 1) and certain power-enhancing modifications are more gradually achieved, or actively embraced, than others.

THE REGIME OF ELIGIBILITY

Aid agencies provide resources for rural development projects that shape peasants into eligibility for modernity, train them for economic viability, and encapsulate them in a development process not of their own making (Verhagen 1987; see chapter 12). The agencies pay too, as local regimes of truth vie with them for the power to pronounce the path to authentic development:

> Nothing I studied prepared me for this. . . . There are so many centers of power – too many chiefs. We thought that if we cleared things with the appropriate ministries OK. But if the chief is against it or the Women's League, or district councillors, there's a problem. Each area also has a chair of the village committee, a ward development committee, a school committee, a local ZANU–PF chair, a deputy security secretary. By the time you are through with all their meetings there is little time left to talk about water holes, for example. You must massage many egos and you have to handle it carefully.
>
> (official of Partners for Productivity)

Many such organizations claim to have "women" clearly in their project sights. Others, typified by the local parastatal Small Enterprises

Development Corporation (SEDCO), are ambivalent on gender issues: "men are rejecting community development projects in rural areas because they associate them with women's development, but women do have difficulties." Some organizations are hostile to projects geared specifically for people called women. They say: "It's not good for women to form their own cooperatives because that's feminism and it causes divisiveness" (official of the Organization of Collective Cooperatives of Zimbabwe (OCCZIM)); or, "we haven't women's programs *per se*, we have community programs and the majority of people we work with are women" (official from Christian Care). Eligibility, it seems, interlocks at various points with the constraints and inducements of other identity-shaping regimes.

THE REGIME OF GENDER

A multifaceted gender regime permeates all the others. It combines precolonial patriarchal practices with the sexual division of labor the Rhodesians introduced, and with many of the gender issues that are encrusted in Zimbabwe's diverse politics of truth today. Viewed benignly, the gender regime "merely" enforces a commonplace designation of two major types of people – men and women. But this designation is rarely benign in its effects on people classified as women. Moreover the regime of gender, despite many historical permutations and challenges – not the least by "women guerrillas" during Zimbabwe's liberation struggle – continues to be openly advocated in the sense that many people defend "traditional" gender distinctions as a true way of identifying all people and their social places. More than the other regimes, this one tends to cross race, class, region, ethnicity and religion, albeit not in a uniform way.

A.K.H. Weinrich (1979) characterized the precolonial gender regime as a lineage mode of production that enabled economic "man" to appropriate the labor and surpluses of biological-economic "woman" through the mechanism of brideprice.[8] Indeed, one can argue that *lobola* (brideprice) produced and reinforced gender by inscribing bodies of community members with "truthful" natures and duties according to their anatomical characteristics. "Women" were the ones whose parents were given gifts to seal a contract that would transfer their valued labor to husbands. Shona "women" then experienced the status of outsiders in the lineages of their husbands and, in addition, were given use rights to land only; the "men" had primary rights (Batezat and Mwalo 1989). In still designating it as true that biological women's valued production should be deeded to a gender-privileged person, the double message of *lobola* trumpets "women" as so powerful that their identity must be controlled by "men."

Rhodesian governments reinforced local customs of male control to safeguard themselves from "native" uprisings, even in cases where they found local gender practices distasteful:

> While colonial officials were highly critical of bridewealth transfers and polygamy, these practices were not considered so offensive that they warranted intervention and the provocation of the older men's ire . . . the native commissioner of the Marandellas District acknowledged that "any legislation to improve the legal status of the woman" within marriage "must necessarily cut at the very root of Native institutions."
>
> (Schmidt 1990: 628)

Some rural "women" found the ambivalences wide enough to permit an "escape" to missions or cities. But by the 1940s:

> due to poor land and climatic conditions of most communal areas, lack of technological inputs and labour constraints, women's food production efforts resulted in gradually decreasing yields especially in the latter part of the colonial era. This made women and children increasingly dependent on the wages of the men for food and other basics. This dependency together with the prestigious earning capacity of the man had a negative effect on women's position and self-image.
>
> (Ministry of Community and Cooperative Development and Women's Affairs, Zimbabwe, Government of 1983: 3–4)

The regime of gender is still evident in the ongoing practice of *lobola*, in customs specifying that women should serve men in the household, and now in letters to newspapers complaining about disruptive women (see Sylvester 1991b, chapters 4 and 6). It animates debates about income-generating projects for "women," as opposed to those that would be for all the "people," and weaves in and out of Marxist allusions to "women" as "mothers" instead of "workers," the "gender blindness" of the liberal pragmatic regime, and the concerns with "prostitution" that rise up in the power relations of indispensability/dispensability. This gender politics of truth does not subordinate all people called women in Zimbabwe to the same degree: "childless African women" are scorned while "mothers" have social standing; "white women" have been long accustomed to privilege *vis-à-vis* African counterparts; there are now laws on the books enabling "African women" to be legal adults at the age of eighteen, and so on. Nonetheless, the gender regime is strong enough to ensure that, for the time being at least, it is widely accepted that people called men are kings of castles and kraals.

ALTERNATIVE REGIMES OF TRUTH?

Where regimes of truth are multiple, overlapping and inconsistent, subjects can mediate power and knowledge at various margins and perhaps find "alternative rules of use that assume alternative ways of life . . . rather than affirming an official ideology by accepting ordinary use" (Shapiro 1981: 107). How do rural "women" who experience contradictory constraints on their identities contest, resist, and acquiesce, and to which regimes of truth do they respond?

In 1988 I interviewed 375 "women" (and 40 "men") in rural cooperatives and farming groups in Mashonaland East Province about their identities, approaches to work, and gender relations in the workplace. At first glance, the two genders were determinant and fixed, readily labeled and differentiated by the respondents. The boundaries between commonplace understandings of men and women, however, tended to blur as "women" expressed the desire to have the power attributes of "men," and as they tried to get those attributes in ways that carved out spaces in-between assigned genders and in-between farming and cooperative sites of production.

Middle spaces were also created by the politics of research relations in the study. Respondents rarely presented themselves to me as individuals – as so much social science survey research assumes and demands – but gathered in groups, as though "views" were legitimate only to the degree that they were collectively generated. The groups carefully controlled the English they used to convey those views and often contested among themselves in Shona the degree to which an English summary of the discussion was "wrongly put" or "too simple." The power of the collectivity was also evident as members repeatedly demanded that I share my "views" with them on the subject at hand, asking: What are women peasants like in the United States"? "What do you think of what we're saying to you"? Their collective insistence was hard to ignore. By not refusing to respond, my knowledge openly entered the "inters" of the "views" instead of surreptitiously figuring into the research as "scientific detachment." The results defy social science regimes of truth-finding and reveal rejections, complicities and amalgamation of regime rules about gender-appropriate work. They also produce "women agricultural producers" in ways that challenge the grouping logic of government planners.

"WOMEN" IN FARMER GROUPS

"Women" in the Agritex groups I studied described their projects as miniature regimes of strength. Each group displayed the "approved" chain of command – "chair, vice, treasurer, vice, secretary" – and members narrated their discipline in similar ways:

1 We have five working units that travel, rotate and program the work. We do mostly planting and weeding. Each person works only for one group. If any farmer has a poor crop, the group can monitor the reasons for this and decide to continue helping or not.
2 We have working units of up to six people. We do almost everything together and even sometimes share draft power. We buy inputs and each working unit markets products together. Individuals also do gardening which they market as a group to keep transport costs low. We also dig as a group. There is a working roster and every member has a day to participate.
3 Our working units rotate from land to land doing all the tasks from plowing to harvesting to herding cattle. We also make compost and dig fishponds for individual farmers. The group has by-laws and each new member pays Z$5 to join.

For their efforts, these "women" reported high proceeds but low net incomes:

1 We're likely to have sixty bags of maize because this is a good year. We'll use fifteen bags for our own consumption and sell the rest at about Z$18 per bag. This will give us Z$810.00. We will also sell some groundnuts and some garden vegetables to bring the family income to about Z$1200 for the year.
2 I grow squash, peas, baby marrow, tomatoes, onions and get Z$500 net off a one acre garden. Most of us make about Z$1,000 a year on maize we sell to the GMB [Grain Marketing Board] – our husbands control that money – and almost all have AFC loans we repay at about Z$600 for three acres. That leaves us with only about Z$400. We can also make thirty or forty cents a day gardening; bigger profits are only for those who have big gardens.
3 Most of our income comes from market gardening. This is what we use for soap and for our own use. We also get between Z$700–800 from the grain marketing board for maize and Z$200–300 for groundnuts, assuming seven acres, and between Z$2,000 and Z$3,000 for maize on eight acres. Very few of us have working husbands and those who do add between Z$30 and Z$180 to the household. We women don't see that income. Those who have loans to the AFC pay Z$600 a year for three acres and at the end of the year have only Z$200 left, even if we make Z$2,000 on our crops.

It has been said that government reform of "native areas" led to a so-called boom in peasant agriculture that may have only affected "at most 20 per cent of the peasantry" (Moyo 1986: 188). Comments offered by "women" farmers in Mashonaland shed light on a political economy of rural profit and debt in Zimbabwe.

The "women," however, spoke highly of the group's influence on their fortunes, saying we can now "help ourselves by rotating work and doing things collectively," or relating how a "problem getting fertilizer and other inputs as individuals" led them to send "two kraal heads to talk to the District Commissioner about forming groups and the DC sent Agritex." The point of reference was one's own relatively poorer past and not some strict economic criterion of success.

Agencies that provided training programs received uniformly favorable reviews, irrespective of the type of knowledge their courses imparted. Specialists in the Ministry of Community Development often concentrated on teaching knitting, sewing and cooking, skills that would seem to reinforce a preexisting sexual division of labor. Less "women"-identified agencies, like the National Farmer's Union, broadened the categories of "women's" work by teaching leadership and bookkeeping. In either case, the farming "women" were enthusiastic and told me that "men say they married us to keep us here working [but] . . . we speak to our husbands and attend training sessions anyway." Viewed as nongendered in their agendas, the helping agencies seemingly gave "women" permission to do something forbidden: to "attend training sessions" in violation of the usual "rule" of the gender regime that "women" may only do what the local "men" allow. That is, the agencies bestowed on "women" a newly *legitimate* identity as "women workers" or "women producers" – something more than "just women." As they took advantage of training, the "women" took on much of the work in the groups except official leadership functions. This imbalance appeared to be acceptable to the "women," as though one needed full training in knowledge located outside the close-by gender regime before one had the resources to challenge the "men." Thus two of the three groups I visited were chaired by "men," even though "women" outnumbered "men" in each group by a large margin and even though "women" were endeavoring in other ways to train outside the usual notion of dependent "women."

There were differences, however, among the "women" I spoke to on the subject of working with "men." Farm groups chaired by "men" relayed the sense that "some of the work needs men" and "men are usually free to go to meetings." The one group chaired by a "woman" expressed a preference for working mostly with "women, because men try to oppress us and acquire power and sometimes our interests differ," as when "women want a knitting machine and the men would be opposed." This difference echoed a tendency I found among "women" in urban cooperatives, who claimed the separate terrain that most regimes assigned them and used it to develop autonomous priorities

(Sylvester 1991a). Whether "men" were requisite in labor, therefore, seemed to depend on the circle one frequented. "Women" working with "women" authorized themselves as legitimate "workers," irrespective of whether they were engaged in training courses that might impart that knowledge. "Women" working with "men" tended to authorize the "men" as indispensables in a rehearsal of gender propriety that may have hidden their agency-training to be more than "just women" or may have reflected a sense that "women producers" are still women and still need men to complement the process.

It initially appeared that "women" in groups chaired by "men" also had a different perspective on the ways "women" and "men" work. One group registered the sense I often heard that:

> Men and women work the same, but we don't agree on things. Women sometimes work alone because the men are drunk. This means our working hours are quite different. Sometimes the husband doesn't go to the field. We are oppressed.[9]

But the "women" were not reconciled to these discrepancies. When "men" failed to go to the fields or, in the words of one "women"-led group, when "some are not prepared to take loans from the AFC," these refusals of discipline constituted negative power that annoyed and burdened "women." Conforming to schedules and taking the initiative in procuring loans – in both cases taking on board a rational purposive discourse of regimentation through training – emerged as measures of a progressive attitude toward work. If "women" could set the standards of communal area farming, they would even out the gender-weighted sense of indispensability by focusing more on rules everyone would follow. Said a consultant for a Food and Agriculture Organization (FAO) project on "women farmers": "They want rules so they know what to do if people don't come to work." Such rules would constitute the positive power of equality lacking in the juridical equality the liberal-pragmatic side of government had granted "women."

Stories of household politics indicated that such equality was already being surreptitiously scripted:

> At the moment we women market to the GMB or to approved buyers here at a lower price. The women don't usually report the latter earnings to their husbands – those earnings are for us. If we need to pay school fees we will sell more bags outside the GMB and pay those fees. If there are no problems in the household, we'll sell fewer bags to outside buyers and more to the GMB.

At the same time, most "women" I talked to worriedly described powerful sanctions that could be wielded against such refusals of proper gender place:

1 We have GMB cards but he'll sometimes take them and use his card.
2 We don't want more children and have arguments with our husbands about this. They threaten us with another wife.
3 We elect men to offices in the group out of fear. . . . Some women don't want to be elected because their husbands are quite against it.

When I asked directly what they would seek to accomplish if they had the type of power usually associated with extraordinary "men" – if they could be the president of Zimbabwe – there was little hesitation in expressing how a community of women's solidarity could push beyond "training" and enter "leadership":

1 A woman at the top would look after women. We would make males go for family planning. We would try to reduce school fees – this is most important, and since the prices of all commodities are going up we would have some review of this. We would give help to groups.
2 If a woman were president of Zimbabwe there would be changes and men's attitudes would change. But the problem is that leaders are supported by men too, so there might not be many changes. . . . We don't know where to start to get power. If we did we would go ahead.
3 If I were president of Zimbabwe I would lower prices of everything, because at the present time it is very difficult for us. The returns we get are quite low. If a schoolboy got a schoolgirl pregnant I would force them to marry. Now men aren't paying for impregnations. I would do something about the community problem of criminals getting acquitted and of people accused of small crimes being punished severely.

The picture these "women farmers" painted looks as complex and cross-cutting as the "ideological" picture in Zimbabwe at large. Yet they said, in effect with one voice, that the gender regime affects them most. Male relatives perpetuate the regime; it is they who can say what counts as true, and they distinguish true from false statements on the basis of the bodies of the speaker. If "women" could gain power, they envision a community of gender solidarity that would mimic and yet reverse the rules of indispensability that surround them in order to bring "true" justice to "women" and their societies. They would enforce equality through a counterregime of gender strength (e.g. "we would make males go for family planning"). "True" statements would then be distinguished from the "false" by rules of unified, socially responsible women's standpoint posed in response to a regime perceived as currently united around men's self-promoting standpoints, albeit there was some concern that not even a true women's regime of strength could cope with unreconstituted men.

"WOMEN" IN AGRICULTURAL COOPERATIVES

Farming cooperatives were also regimented, but varieties in discipline and hierarchy greatly exceeded allowable parameters scripted in official cooperative guidelines, as this selection of comments indicates:

1 We have a chair, vice chair, secretary, treasurer and three committees: management, which runs the business; supervisory, which oversees the physical work; and the grain committee, which buys grain from members and sells it to the GMB. Elections are held every year and there is a general meeting once a month.

2 At this collective there are two leadership structures: for the men and for the women who are not yet members. Each has a chair, vice, secretary, treasurer and two chief committee members responsible for all activities including entertainment. There is a production manager who coordinates both groups. The men and women work together, but because women don't have a joining fee paid, they're separate.

3 The cooperative is organized around an executive committee of the chair, vice, secretary, vice, treasurer, all of whom have held their posts since 1981 or 1983. There are no committee members. Elections are every two years with a general meeting once a month.

4 Seven out of the ten members are on the executive committee.

The organizational variation suggests that some Zimbabwean agricultural cooperatives are more in step with the regime of pragmatism, and with the notion of pluralist organization, than with the politics of a promised Marxist order – "you tell us what socialism is because we have no idea" – or the conformist politics of strength.

All the cooperatives I visited grew and marketed maize, groundnuts and vegetables, and some also had poultry and their own shops. The collective cooperatives lived off their profits and the other cooperatives usually supplemented farm group incomes. The profits varied as much as the management structures, but not more than the profits "women" in farming groups reported:

1 We owe the AFC Z$11,000 on a loan of Z$38,000. It will never be OK because we owe so much. We have three boreholes on this fifty-one hectare farm but they are not running and . . . companies say coops don't pay back, so even though we have budgeted Z$200 to fix one of them, the companies won't come. Also every year we have to hire a tractor to plow and we don't have money for this now (collective cooperative in Goromonzi started in 1985).

2 We are running the coop on a red line all the time. It's not easy for coops to produce . . . It's a crisis sort of experience. We've been borrowing from the AFC for fertilizer, chemicals, etc. They have

treated us well and we have medium and short-term loans. We pay back Z$155,000 this year and make Z$200,000 (collective cooperative in Seke started in 1981).

3 We have Z$700 in the bank. The average share per member per year is Z$60 (cooperative in Seke started in 1982).

4 We sell from the fields once a year and if there is surplus we share up to Z$25 a month (collective cooperative in Marondera started in 1982).

5 We have never had any money to share (cooperative in Chiota started in 1981).

6 We have shared money twice in the existence of the cooperative – Z$10 per member each time. But we have more than Z$592 in the bank so that when we think of another project we can add this money for it (cooperative in Chiota registered in 1983).

7 During the period of high demand for fertilizer we make Z$1,000 to 3,000 a week, and net about Z$17,000 a month . . . So far we have only been banking the money and have not given anyone pay (cooperative in Charare village started in 1982).

Service and aid organizations initially jumped on the bandwagon of cooperative production and earmarked some funds for these new ventures. However, when cooperatives did not flourish by the late 1980s, donors often questioned the viability of this entire area of production. As the comments above show, "women" involved in cooperative production reported little that would suggest that cooperatives were usually doing well as businesses. And yet "women cooperators," no less than those involved in farm groups, proclaimed their struggling enterprises "successes":

1 We have satisfied customers and are locally supplying what people need. Formerly we waited for the AFC to give us loans and we could be rejected. We used to walk to Murewa to sell two bags of maize for money. Now I just go to the coop and sell my grain and it's cheaper for me (Charare).

2 Things are OK and improving. We make a bit more money each year. Given that we are only twenty-two members we are satisfied (Matoko).

3 We're a success when compared to other cooperatives (Chiota, Seke).

4 This cooperative was a success until the drought. The problem is how to get capital to start a garden again (Chiota).

5 We do any work which needs to be done . . . knowing what we do now we would start a cooperative again (Goromonzi).

These cooperators measured viability in dollar terms *and* in terms of how much better off their group seemed compared to other cooperatives they knew. To be "successful" in this sense was to be viable in unconventional ways.

Cooperators reported varied contacts with the purveyors of aid. One collective cooperative was into the fifth year of funding by an international donor and the question loomed of "viable" life after aid. The cooperative in Charare village reported that "our operation needs reasonable capital to serve its members and also more training. We looked for money but couldn't get much." Another group of cooperators approached SEDCO for a loan to start a grinding mill and were told to raise Z$800 themselves first; when they did this SEDCO told them to (discipline yourselves more and) raise Z$2,000. The remaining cooperatives reported dribs and drabs of money from donors. The (male) chair of the collective cooperative in Seke said bluntly: "We'd run for donor money if it were there, but we can't go and say we need seeds, fertilizers and other everyday things because they don't give for those." Agritex received praise from two cooperatives for helping them to obtain land, for doing "a five year plan, and for tr[ying] to get people from Coops to come out." But cooperatives that described themselves as "successful" were by no means amply and steadily supported by donors.

Silveira House, a local Roman Catholic organization with a long and more consistent history of outreach to communal areas in Mashonaland East, hosted valued courses in leadership, retail marketing, cooperative management and bookkeeping. As I observed several of these sessions, I wondered who among the cooperators were being trained. The lecturers (at that time all "men") tended to address the "men" in the audience more than the "women," and they took more time with "men's" questions. No reference was made to gender imbalances in cooperative committee structures, the possibility of a sexual division of labor in a cooperative, or to "women's" dual work in support of families and cooperatives. These absences may have been unimportant to the "women" attending the sessions, for whom simply being there, despite being not-men, was a plus. But the regime of gender was surreptitiously at work within the regime of eligibility, such that "women" were sometimes ghosts in the classrooms.

They were less ghostly in day-to-day relations in the cooperatives. With the exception of cooperatives set up by male ex-combatants in the years immediately following independence, all showed the same rural demographic tendency to attract more females than males (or in one case an equal number of each). This is a legacy of colonial labor policies that reserved jobs for "men" in towns. Four cooperatives had "women" chairs and four entrusted "men" with leadership positions. In either case, most "women" cooperators expressed preference for working side by side with (the unproblematized group called) men rather than working exclusively with (given) women, mainly because "certain tasks need men" and "certain tasks should be done by women" (e.g. making soap, sewing and knitting). It would invariably come out, however, that "our husbands have power but women work harder – we work six hours and men work eight hours, but some women get more done." The gender regime was doing battle with the liberal-pragmatic truth that "women" can work alongside "men" in the new Zimbabwe.

Often, "men" felt entitled to tell "women" that their work performances were below standards:

> The women had no experience of running a cooperative. So men will join and give them that experience. Up to now we haven't joined because we wanted to see how they were doing. We see they're failing and some will join.

This audacious statement came from a "man" who himself had no prior experience working in a cooperative. In another case at a showcase collective farm, several "women" complained of being barred by their husbands from formal membership in the cooperative. The wives of several cooperative committee members were among the marginalized; indeed, the husband of one of them was the chair of the cooperative and of the national executive body of OCCZIM. I remembered that the Secretary General of OCCZIM had told me that "collective cooperatives mix and do not divide people . . . the major principle of cooperatives is open and voluntary membership." In such cases, the unresolved "Woman Question" in Marxist regimes of promise conjoined with a gendered indispensability/dispensability division promoted by the regime of strength, and with the pragmatic tendency to leave certain things (traditional gender roles) alone.

Even in cooperatives where discrimination against "women" was condemned, "men" could often say what counted as true and they pronounced it true that "women" should be equal to "men" in cooperatives. One male chair had an intellectual way of disavowing old gender truths while distancing himself from the realities of inequality on his cooperative:

> We have forty-three members and I don't know how many are women. It has to be like that because if some members in a coop are going to be women, from my point of view, then there must be a purpose to that. These are the members and this is what we do. I don't think male and female except outside the coop . . . I don't want to think women are there in the creche because they are women. I don't want to think I'm chair because I'm a man.

"Women" in cooperatives, by contrast, generally told me:

199

1 Zimbabwean women want more power. We are as good as men. With extra power we can work hard to show men we're just the same.
2 If given power we'd like to know how the money in this cooperative goes in and out. We aren't well enough educated to know this.
3 We want skills. We want to drive the tractor and the car.
4 We're not power hungry. To each according to ability and for the betterment of the project.
5 With power we could work harder.
6 Women want to be treated equally and if assigned a position want to be respected and treated as equals. They also want to be encouraged and need much encouragement.

In these remarks, there is a sense that in order to rectify gender problems in the workplace, "women" generally need "extra power." With it, "women" would institute sweeping changes:

1 Prices for our crops would rise because women know that women work hard on the crops and don't get enough money. We'd do something to improve the prices for women while men are in town drinking.
2 We'd change the manpower balance in the Ministry of Local Government. There are too many men there and when a woman goes looking for a birth certificate she's given a rough time. Also if women were in the Ministry of Education they would do something about the high school fees.
3 We'd have more education for everyone. We'd also give a chance to those who can't do other businesses and we'd give coops loans. We'd create more jobs and the qualifications to do them. We'd import machines and send people out to other countries to see what they're doing.
4 We'd have a woman president and bring to her all the problems in our clubs, families, and coops, like having funds to buy fences and a grinding mill. She would not impose her power but could bring power to Chiota. Of course it is also possible that nothing would change if a woman were president.
5 If women needed courses we could send someone without asking our husbands.
6 We would develop our country and try to give more to women, get cars, supermarkets, my own business, shops, cattle, everything.
7 If there were a woman leading the country there would have to be men near her to help her.

How to get coveted resources without an outside grant of power was a mystery to many of these "women." However, there was little hesitation on their part that they could handle the same types of power that the gender-privileged group enjoyed. They wanted power, not only "to show men we're just the same," not only to begin to be heard, seen, and appreciated as partners in work and cooperative management. They also wanted power to alter the social agenda by trying "to give more to women." Here there is evidence that rural farming "women" in Mashonaland would, if they could, replace what they see as restrictive gender power with socially expansive and inclusive power.

Overall, the cooperatives were as preoccupied with managing, accommodating and changing the regime of gender as were the Agritex farming groups. The state and its regimes were not problematic. Donors did not stand in the way. Local "men" were the problem. "Women" tried to train themselves into comparable gender competence under the apparent assumption that without "indispensable" and male-linked skills, "women" and their enterprises could be pronounced "unviable" as "self-sufficient" "producing" identities (see chapter 12). Thus, in the face of poor economic performances by their cooperatives, the "women" members expressed appreciation for their enterprises and evidenced determination to train for a viable sense of "successful-women-producers" working equitably with "men."

WHERE'S THE TRUTH?

Zimbabwean "women" in agricultural groups face a host of regime representatives in the form of male relatives, mouthpieces for overlapping state regimes of truth, representatives of aid regimes, and researchers like myself who momentarily take them away from their work in order to listen to their voices (see Spivak 1988). They do not respond with a uniform strategy. "Women" in farming groups seem to replicate the truths associated with the regime of strength by speaking wistfully of turning the tables on many of their masters through a series of bold insubordinations. "Women" in cooperatives take the pragmatic road: they agree to follow official stipulations but set up shop as they please. Both groups take any education provided and hope for the (extra) power that can set their social agendas in operation. Both also focus resentments on the politics of gender. No one questions gender as a meaningful identity: there are men and there are women and everyone knows who is who. The issue is which identity group should have power and not whether gender groupings make sense.

Similar views about the gender regime reappear across work contexts and suggest that these "women" see power as something one accumulates socially; it is not a fixed biological birthright. The road to accumulation of power for "women" requires that extra boost of education and skills that comes from training with outside experts – even if those experts do not see you in the classroom. Drinkwater (1991) maintains that rural Zimbabweans resist the technocratic lifeworld and seek instead to install a very different knowledge-power regime of "the peasant." The "women" among the peasantry I met, however, acted as though any different regime the local "men" ran would be more problematic than the "proven" one the Rhodesians had brought and the government now promoted. If the "women" I interviewed could, they would retain the power of gender, augment it with education, link it to a "women's agenda," and exercise it for new purposes. In effect, they pitted a lifeworld of local gender rules that disadvantaged them against the lifeworld of purposive rationality, of modernity, of outsider knowledge that, to them, opened the doors to power. They regarded outside expertise more as leverage in their gender struggles than as patriarchal structures designed to perpetuate gender hierarchy (see chapter 12).

Gaps between aspirations and daily realities lead us to question whether everyday struggles (Scott 1985) by Mashonaland "peasants" who are "women" have the potential for not-so-everyday reconstitution of androcentric regimes. There are spaces in, around, and between regimes of truth for "women" to narrate common and differential experiences with and against usual authorities. One must not, however, refuse their experiences of everyday powerlessness and conclude that these subjects are establishing alternative regimes of truth. To do so would be to stretch their experiences too far. One might say that they are reshaping themselves in small degrees, in part by aligning with sympathetic donors who symbolize that extra power required just to "induce the regular effects of power" if one is "woman." Nonetheless, their efforts to tug at the parameters of gender reveal the beginnings of many potentially viable "truths" about "women agricultural producers." We will not be able to see and appreciate those tugs as long as we call certain people women without thinking twice about what that means for the subjects so named, for the series of power relations that swirl around them, and for the production programs in which they participate.

NOTES

1 I wish to thank the Economics Department of the University of Zimbabwe for facilitating this research, the "women" I interviewed in Zimbabwe for relating their stories, and participants at the Workshop on Gender on the Frontline, held at Northwestern University in June 1991, for useful critiques of a first draft.

2 It is not easy to differentiate traditional practices from the traditions colonials invented. Moreover, any search for authenticity can lead one into an impenetrable maze. For a sense of invented traditions in Zimbabwe, see Ranger (1985) and Garlake (1982).

3 Donna Pankhurst (1991: 611–12), for example, tells us "to incorporate an

analysis of gender into any future formulae that attempt to improve agricultural practices or to change land tenure in any way." She does not, however, question the pre-given categories of men and women.

4 There are many other terms and identities we could problematize in this study, including "farm group," "cooperative," "Zimbabwe" and "rural." All are socially constituted statuses with boundaries that demarcate the inside from the outside, and all the boundaries are porous enough in practice to enable transgressing migrations. This chapter is designed to illuminate the themes of the volume and focuses, therefore, on problematizing women as "women." (When I refer to literatures or local views that do not problematize "women," I do not use inverted commas.) One should bear in mind the other artifactual categories.

5 This may change in the future owing to recent legal moves that enable the government to seize underutilized or unoccupied lands for resale or resettlement.

6 To appreciate what happens to "prostitutes" in Zimbabwe, see Weiss (1986). I am grateful to Jane Parpart for pointing out that the regime of strength resonates with patriarchy. That is, it can be understood as a hierarchy of power relations based on the presumed indispensability of "men" to social progress and order, thus blurring boundaries with the regime of gender. But there is a difference: the regime of strength addresses and makes "function as true" many power relations of indispensability/dispensability that can affect *"men" too*, particularly "men" who challenge ZANU–PF. For a discussion of ZANU–PF's efforts to impose the regime of strength on Edgar Tekere, who "deigned" to compete with Robert Mugabe in the 1990 elections, for example, see Sylvester (1990a).

7 Most Zimbabweans I interviewed felt "the government could be doing more for us," but were unhappy about new policies and programs that would disrupt their usual habits of production, labor and domicile. Those who hoped to be resettled through land reform projects, for instance, wanted to be in villages with minimal common production and marketing activities. Radical solutions, such as collective cooperatives, were seen as too extreme.

8 There is a lively debate in Zimbabwe on the pros and cons of brideprice. For an overview, see Sylvester (1991b, chapter 5).

9 Pankhurst (1991: 625) notes that "[i]t is now socially unacceptable to complain publicly about one's [own] husband, and even women who have suffered in the most extreme ways do not do this"; thus, perhaps, the references here to "the husband" or to "men" in general.

11

GENDER AND DEVELOPMENT IN ACTION

Eva M. Rathgeber

In the 1970s and 1980s, the term "gender" increasingly began to be used in (English language) discussions of the different roles of women and men in social processes.[1] "Gender" was defined as a socially constructed category that carries with it expectations and responsibilities that are not biologically determined. By the mid- to late 1980s, the gender concept also had become common in writing on development issues and in some circles it replaced the earlier Women in Development (WID) approach. While this move to replace "women" with the more neutral term "gender" has been and continues to be criticized by some feminist analysts of development processes, this chapter argues that the gender approach avoids the pitfalls of economic determinism inevitably linked with the earlier WID approach. Along the lines of the postmodernist critiques discussed by the editors in this volume, the Gender and Development (GAD) approach moves beyond the instrumentalist definitions offered by the earliest Women in Development theorists who attempted to fit women, and most specifically developing country women, into predetermined categories which themselves were based on essentially linear, progressivist, Western views of "modernization."

Although the gender concept is broader in scope than WID and at first glance may appear to be less threatening to the status quo, it ultimately provides a deeper analysis and makes demands for changes that will effect more profoundly the structure of current social, economic and cultural processes. As will be discussed below, when taken to its logical end, the GAD approach rejects the very foundations upon which international development assistance programs have been structured. At the same time, however, the implementation of development projects grounded in gender analysis, and indeed even the *conceptualization* of their implementation, has proved to be extremely difficult.

The discourse of liberal development thought has changed periodically during the past thirty years but, to a significant extent, the major development strategies have been based on similar assumptions. Some of the key phases of liberal development thinking since the 1960s, include

(1) human capital development (1960s); (2) technological development (1960s/early 1970s); (3) basic needs (mid- to late 1970s); (4) Women in Development (1980s); (5) structural adjustment policies (SAPs) (early 1980s to present); and (6) sustainable development (late 1980s to present). In each case, the failures of earlier strategies have led to the establishment of new approaches.

For the most part, new approaches have arisen out of an essentially separatist mentality. Usually a particular issue or set of issues has been identified and removed from its social context, first for specific economic analysis and ultimately for donor agency and NGO action and support. Most frequently, "new" approaches have been proposed in light of the perceived failures of earlier efforts to improve overall economic performance and, by implication, standards of living in developing countries. Usually, such approaches have been based on an essentially economically-based and Eurocentric view of what constitutes an acceptable "standard of living." Significantly too, while new approaches have usually been formulated to more explicitly address the "real needs or interests" of developing country people, they have rarely if ever originated from the supposed beneficiaries themselves. Inevitably they have been interpreted by academic or practitioner intermediaries who see themselves as experts on the South and on development. This in itself provides insight into the relative lack of attention given to opinions and experiences of others. The alliance of academicians and practitioners has effectively created a world of development "insiders" and "outsiders" and curtailed opportunities for alternative voices to be heard and to form a basis for action. The great irony of course, is that the "outsiders" are the beneficiaries themselves.

Rarely, if ever, have development efforts been holistic in conceptualization. Inevitably, different interest groups, spanning both the scholarly and the donor/NGO communities, champion the cause of each "new" approach to development.[2] They form alliances which enable the adoption of core sets of ideas and methodologies that become the central knowledge and practice of the "new approach." Consequently, there tends to be very little learning from earlier experiences. For example, few of the mainstream agencies working on environment today are factoring into their analyses knowledge and experience gained in the 1980s on the sexual division of labor in agriculture, the different relationships men and women have with natural resource bases and the impact of male migration on women's work and responsibilities in many parts of the world. The tendency instead is to follow more or less the same path as in the past: i.e., (1) identification of environmental problems, primarily from a macro perspective; (2) identification of economically efficient approaches to redress the problems; and (3) implementation of technologically-based solutions. In this sense, development research and practice tends to constantly rediscover things already known.

This chapter will discuss some insights that have emerged from both theoretical and practical work done by practitioners working within the Women in Development and, later, Gender and Development approaches during the 1980s. It will point to important similarities and differences between the two and then will examine how these insights could be brought into project design in a contemporary global context when women's problems and concerns in general are receiving less attention from development agencies. Most significantly, the chapter examines strategies for the more effective incorporation of developing country women's own definitions of themselves and their perspectives/knowledge(s) into development discourse as well as program design and implementation.

A GENDER AND DEVELOPMENT OVERVIEW

The Gender and Development approach can be seen as a predictable analytical outcome of the earlier WID approach. As we have seen, WID put emphasis on providing women with opportunities to participate in male-defined and male-dominated social and economic structures, while GAD questions the assumptions implicit in these structures. It sees women as agents of change rather than merely as recipients of development assistance, and stresses the need for women to organize themselves for effective political voice (Rathgeber 1990). It recognizes the importance of both class solidarities and class distinctions but argues that the ideology of patriarchy operates within and across classes to the disadvantage of women. Consequently scholars working within the GAD perspective have explored both the connections among and the contradictions of gender, class, race and development. Thus a Gender and Development perspective leads to a fundamental re-examination of social structures and institutions, to a rethinking of hierarchical gender relations and ultimately to the loss of power of entrenched elites, which will effect some women as well as men. It focuses on both the *condition* of women, i.e. their material state in terms of education, access to credit, technology, health status, legal status, etc., and the *position* of women, ie. the more intangible factors inherent in the social relations of gender and in the relations of power between men and women (Young 1988). These two concepts parallel Maxine Molyneux's (1985) distinction between women's *strategic* gender interests (their position) which focus on the analysis of their subordination, and their *practical* gender interests (condition), which are concerned with alleviation of specific and concrete disadvantages faced by women (see chapter 3). It is clear that both are important.

206

From a postmodern feminist perspective, one of the most significant underlying assumptions of the GAD approach is that the situation (i.e. the condition and position) of women is a function of multiple power relationships. There is no single "women's situation." Women's lives are affected by multiple variables such as race, ethnicity or class. For this reason, both in the North and the South there can never be a single voice that expresses all women's concerns or perspectives. Yet, much of the WID and general social programming of development agencies has been formulated on the assumption that there is a single woman's voice, and that "voice" is drawn largely from the experiences of white, middle-class women in the North.

While the "gender" approach became common in critical feminist writing of the 1980s, it had a somewhat slower acceptance in development practice. Even in the 1990s, development agencies that give specific attention to women's needs tend towards WID programs. On this issue, as on some others, the development agencies have tended to lag behind critical thinkers. Jane Parpart and Kathleen Staudt (1989) point out that while "development" was the major preoccupation of writers on Africa in the 1960s and 1970s, "crisis" became the principal theme of the 1980s. Conversely, most development agency programing is still operating within the realm of "development," i.e. capacity-building and institution-building. Perhaps this is not surprising given the fact that most donor agencies, particularly bilateral ones, act primarily as "brokers" of development. They are tightly constrained by the policies of their own governments and have only a limited capacity, or even desire, to influence practice and effect societal changes in the developing countries where they work. The social relations of gender are labeled as falling into the realm of culture and strong advocacy for a rethinking of gender relations would be seen as unwarranted "cultural interference."[3] This reluctance in turn is reinforced by the high proportion of senior male staff in donor agencies, many of whom do not necessarily see the current social relations of gender as fundamentally problematic. It is easier for them to accept a WID approach which argues on grounds of efficiency and equity for female access to resources and decision-making.

However, even while the basic attitude towards gender has remained conservative, in the 1980s "women's issues" were put on the agenda in many development agencies, particularly in the Scandinavian countries, the Netherlands, Canada and the United States. The 1985 Nairobi meeting increased international pressure on development agencies to integrate women's concerns into their programs. *The Nairobi Forward-looking Strategies for the Advancement of Women* (United Nations, 1985), which were finalized at the meeting, provided specific guidelines that governments could follow to ensure better representation of women and women's interests in all of their programs. The document emphasized the need for a women's perspective in human development and it contained concrete proposals for action with respect to employment, health, education, food, water, agriculture, industry, trade and commerce, science and technology, communications, housing and settlement, community development, transport, energy, environment and social services (Andersen and Baud 1987). However, the strategies still tended to see women as an essentially homogenous group. The Nairobi Strategies were adopted by 157 countries and various meeting have discussed their implementation. Pressure from some government donors have encouraged much greater attention to "women's issues" within the UN system. Nonetheless, there have been wide variations in the extent to which concern for women's needs was actually integrated into agency planning and programming.

EXPERIENCES OF SOME DONOR AGENCIES

Irene Tinker (1990) notes that the two major approaches used in the 1970s and early 1980s by development agencies in conceptualizing programs for women were *welfare* and *efficiency*. In each case emphasis focused on only one aspect of women's roles – reproductive in the case of the welfare approach, productive in the case of the efficiency approach. Moreover, as pointed out by Mayra Buvinic (1986), projects for women frequently were started without feasibility studies with the result that they sometimes proved to be unsustainable for the long term. Frequently the lack of experience and the marginalization of the project implementors, who have tended to be situated in NGOs or secondary government departments rather than in mainstream government positions, has been transferred to the project participants, reinforcing a cycle of failure. In other cases as noted by Moser (1989; 1993), projects have been established by government planners under the erroneous assumption that women's needs and modes of interaction are identical to those of men and that they will respond similarly to created opportunities. To a significant extent, the human capital development efforts of donors in the 1970s (concentrating on the provision of training opportunities) most often overlooked women and girls, implicitly accepting the prevailing cultural norms in many countries which gave preferential treatment to men and boys.

Tinker (1990) notes that by the mid-1970s liberal feminists were arguing that it would be desirable, from an economic perspective, to teach women new skills which would enable them to become self-sufficient and to move away from welfare provided by the state or voluntary agencies. This efficiency argument coincided with the ascension of the "basic needs approach" in development planning, which responded to widespread criticism of mainstream development policies. Advocated by the ILO in 1976, by the late 1970s it had become the guiding ideology of many bilateral donor agencies, including the Canadian International Development Agency (CIDA). This new approach cast some doubt on the effectiveness of earlier technology-intensive, large-scale infrastructure projects, and led to a broader consideration and discussion of the integration of women into development and spearheaded an interest in the production of so-called "appropriate technologies" for women. Efficiency has continued to be a dominant criterion in development agency project implementation, however. It continues to be used as an evaluative tool (USAID 1987), although the definitions of "efficiency" and "efficiency measures," have changed over time.

Nuket Kardam (1991) has analyzed the responses of three different agencies – UNDP, the World Bank and the Ford Foundation – to the integration of women and women's concerns into their programs. She concludes that the range of responses has varied from low in UNDP to high in the Ford Foundation, with the World Bank taking the middle road. On the whole, however, women's concerns have been integrated into ongoing projects in an *ad hoc* manner.

1 *United Nations Development Programme.* UNDP began to give minimal attention to women's concerns in the 1970s, due primarily to the work of Ulla Olin (Snyder 1992). Through her efforts, UNDP set up a system for assessing the level of integration of women's needs and concerns into its programs and projects. However, UNDP established a formal Division for Women in Development only in 1986. USAID in contrast, set up its WID office in 1974 and SIDA, the Swedish bilateral agency, began supporting projects aimed specifically at women as early as the 1960s (Rathgeber 1988).

According to Kardam, the relatively late entry of UNDP into systematic WID activities was due to the decentralized nature of decision-making and policy-making within the UN system. The UNDP coordinates development assistance projects in response to requests and priorities articulated by national governments. Kardam argues that the implementation of WID policies has tended to be dependent on the external pressure of gender-sensitive donor governments and the personal interests of UNDP staff members. Canada, Sweden, Norway and the Netherlands have been particularly active in encouraging UNDP projects with a focus on women.

Current thinking in the United Nations system strongly endorses the promotion of "self-reliance" for women. The 1986 *World Survey on the Role of Women in Development* presented material on the role played by women in the global economy (United Nations 1986). The first update in 1989 underscored women's limited economic progress during the decade (Pietila and Vickers 1990). It paid particular attention to the adverse impact of structural adjustment policies on women. However, the *Survey* also focused on the relationship between women's productive and reproductive roles. In this sense therefore, thinking within the UN system also has been moving beyond the crude dichotomization of early WID initiatives.

2 *World Bank.* The World Bank has had an Advisor on Women in Development since 1977, but in the 1987 restructuring of the Bank this office was expanded and given a higher profile. One of its areas of interest has been the "Safe Motherhood" initiative, based on the argument that "improving maternal health helps involve women more effectively in development" (Herz and Measham 1987). Improvements in maternal health have been linked to a decline in birth rates, which is seen as essential to increased economic efficiency and productivity. This then, has been a major focus of Bank activities with respect to women. At the same time, Bank initiatives with a gender focus have tended to assume the homogeneity of women in the South.

Efforts have been made to produce WID country strategies and by 1990 strategies had been produced for seventeen countries. They include in-depth analysis of women's labor market activities and the economic contributions of women in the countries in question. Nonetheless by the late 1980s, inclusion of women's issues in Bank projects was still primarily a reflection of the interests of individual Bank programs and staff members (Kardam 1991). An overall internal evaluation of WID programs did not occur until 1993 (World Bank 1993). This report supports Kardam's assertion that most Bank staff remain convinced that if integration of women's concerns into program delivery will make projects more economically efficient then ultimately they will be included without special efforts or manipulation. She notes that for many Bank economists "efficiency and productivity are seen as value-neutral and objective," while "equity [is seen] as value-laden and subjective" (Kardam 1991: 116).

3 *Ford Foundation.* In contrast, the Ford Foundation has pursued an entirely different strategy but one which in the long run may have been the most effective. The Foundation has recognized the importance of women's issues since the early 1970s and grants for WID projects increased from 52 for 1972–9 to 326 for 1980–8 (Flora 1983; Kardam 1991). As a private voluntary organization, Ford has been able more easily than either UNDP or the World Bank to act as an advocate for the integration of women's issues into general development initiatives. Instead of establishing a specific WID office or program, the Foundation has attempted to mainstream activities WID into all its programs. This has been done with a high level of

consistency accompanied by periodic monitoring to ensure that women actually are benefiting from the Foundation's activities.

4 *Canadian International Development Agency.* CIDA offers a good example of a bilateral agency that has had a strong commitment to the integration of women's concerns into its programs. CIDA established a WID Directorate in 1984 and by the late 1980s all CIDA professional staff systematically underwent training in gender analysis. CIDA's attitude towards gender issues continued to become more sharply focused throughout the 1980s. By 1991 CIDA policy documents stated clearly and uncompromisingly that WID was considered to be a fundamental development issue rather than a "women's issue," and that in order for the situation of women to be improved, it would be necessary for societies to undergo structural changes at the social, political and economic levels. In other words, although CIDA continued to use the language of "Women in Development," the meanings attached to this language encompassed a "Gender and Development" approach (CIDA 1992).

This leads to an interesting contradiction and gives rise to speculation about the discourse of development and the extent to which meanings can be manipulated by progressive individuals working within donor agencies. Indeed, a closer analysis of the policy statements and the actual program implementation at the project level of various donor agencies might uncover considerable variance between official agency positions and their actual work. However, it would be unduly optimistic to assume that such contradictions or variations always will work to the advantage of women.

It would seem that in many agencies, progress towards the implementation of gender-based concerns into programming has to a considerable extent been due to the efforts of progressive individuals. The recruitment of more women into professional jobs in development agencies has also made a difference, although of course not all women working in donor agencies necessarily share a concern for or interest in women's issues. At another level, however, it is worth noting that the discourse on gender issues within agencies has taken different directions. In some cases, agencies have continued to work within the WID framework, officially promoting policies aimed at enhancing women's economic position and giving recognition to the importance of women's contribution to development processes. However, individuals within the agencies have sometimes concerned themselves more specifically with the *position* of women and have tried to develop projects which address issues of social change and gender power relations. In this way, WID language has been used to mean GAD. Less frequently, GAD language has been used to mean WID. On the whole, although most donors have been reluctant to adopt outright the Gender and Development approach, some of its underlying principles seem to have permeated a number of institutions. Fifteen years of programing for women has shown that projects intended simply to "integrate women into development" are not enough. First, as the scope of women's labor has become better understood, it has been recognized that women have always been part of development and consequently that development cannot proceed without their cooperation. Second, structural changes in the essential power relations between men and women are increasingly being seen as a necessary precondition for any development process with long-term sustainability (Palmer 1991; Stamp 1989).

The contradictory nature of discourse within organizations also deserves analysis. While it is true that most development agencies have specific modes of operation and formulae for program delivery, sometimes considerable space exists for personal creativity in the interpretation of those guidelines. Individuals within the agencies often work in the development area by choice and are committed to issues of social justice. The gentle sabotage of agency norms that has been noted is not an accident. It is the expression of strongly held views and should be recognized as such. A truly postmodern feminist approach to development would recognize and build upon the reality of these diverging views rather than denigrating the efforts of development agencies as either inherently conservative and self-serving or excessively bureaucratic. Obviously agency workers have neither the capacity nor the right to speak for the intended beneficiaries of their efforts and the important issue of how to bring in those voices is addressed in the final section of this chapter. However, the discourse of progressive individuals within agencies should be recognized and interpreted as an alternative voice in itself.

SECTORAL APPROACHES

To better understand how GAD concerns can be integrated into development programming, it may be useful to look at some of the gender-based insights that have emerged in examinations of different sectors. Attention will focus on two areas that have received particular attention from both development scholars and practitioners: environment and science and technology.

1 Environment

The linking of gender and environment is still at an early stage but this is clearly an area which allows for both innovative conceptual work and practical interventions. In a recent paper on sustainable development, CIDA recognized a clear connection between gender equity and environment based both on women's close relationship with natural resource bases (trees, water, land) and on the central role of women in economic production (CIDA 1991). Nonetheless, this recognition is still far from generally accepted either by scholars writing on environment and sustainability issues or by development agency personnel. For example, Michael Redclift's otherwise insightful analysis of the relationship between global environmental crisis and capitalist expansion and the linkages between North and South gives only cursory attention to the issue of gender and different attitudes held by men and women towards the environment (1987).

Similarly, most of the work done by the World Bank on environmental issues is totally gender "neutral," which in practice usually translates into a male bias. There is an implicit assumption that the environment and environmental issues are perceived in the same way by males and females. Given the context of men's and women's differing experiences with the environment and their often divergent needs from the natural environment, this assumption may be ill-founded. As Bonnie Kettel (1993) argues, both "gender" and the "environment" are social and cultural constructions. Perceptions of "environment" are thus dependent on human interaction with and use of the natural landscape. From this perspective, perceptions of environment are closely linked to the sexual division of labor, though mediated by cross-cutting factors such as class, ethnicity and race.

An emerging feminist analysis of the environment has drawn attention to significant linkages between women's roles, the global commoditization of agricultural production and environmental degradation (Dankelman and Davidson 1988; Mackenzie 1993). In face of demands for increased agricultural production and higher levels of efficiency, African farmers are being forced to adopt agricultural practices which will have severe implications for the long-term sustainability of their resource bases. Despite ample demonstration that intensified cropping, higher yielding varieties, irrigation and high levels of fertilizer and pesticide use all strip the land of its nutrients, structural adjustment policies advocated by the World Bank and the IMF and put into practice by the national governments of increasing numbers of African countries, have led to the large-scale adoption of ultimately harmful agricultural practices. In the context of donor-driven efforts to promote the utilization of technology packages, existing knowledge, passed down through generations to both women and men, has been marginalized as part of a systematic silencing of "other voices."

Jane Collins (1991) suggests that the literature on women and environment has often treated women as a homogenous group, overlooking differences of class, caste or ethnicity. Consequently it has tended to generalize about the role of women in resource management and to

213

minimize rigorous analysis of how women's interaction with environment has changed over time and in varying situations. In this sense, the literature on environment has not benefited from the earlier insights of scholars writing on the sexual division of labor and responsibility in agriculture (see also Agarwal 1991). Similarly, development agencies have tended to have a slow reaction time. Although the significance of male migration (and increasing female migration) has been acknowledged widely for over a decade, few agencies have reformulated their programming to take into account the constraints faced by women heads of households.

The hesitation to link gender and environment in a systematic way reflects a general tendency for development analysts to strive for clear boundaries among and between concepts, in order better to understand aspects of the whole, and only later to attempt to make explanatory linkages among them. It is easier to begin with an analysis of "environment" and later to consider "gender," than to begin with an analysis of "gender and environment." As a general rule, we have tended to organize different experiences into separate analytical categories. This can be a problem. For example, Kettel argues that perceptions of the environment are closely related to the sexual division of labor, and that since much of women's environmental knowledge is associated with their agricultural and forest work, which is accorded a low status, their knowledge has also been undervalued. Significantly, the components are not equal to the whole. An analysis of women's agricultural work does not yield a thorough understanding of the nature of women's relationship with the environment and the depth and breadth of their knowledge of environmental issues. Similarly, a focus on women's environmental knowledge *per se* may overlook or undervalue the complexities of the work tasks themselves and the extent to which knowledge is closely linked to them (Kettel 1993). Elizabeth Ardayfio-Schandorf's work in northern Ghana reveals the close link between women's knowledge of forest fruits and vegetables and their responsibility for gathering forest resources (1993). However this instrumentalist knowledge is situated within what might be termed women's "meta-knowledge" of forest resources, which includes both natural science as well as cultural and symbolic knowledge. It is this "meta-knowledge" which guides women in their actual day-to-day decision-making. However, the narrow definition of knowledge, particularly scientific knowledge, adopted by most development specialists, undermines serious consideration of this meta-knowledge, and consequently impoverishes both development thinking and practice.

Melissa Leach (1991) notes that the "women and environment" approach has many of the same weaknesses that have already been identified in the WID approach. There is a tendency to attribute special responsibilities to women based on their "close relationship" with nature which in turn is based primarily on their domestic responsibilities (gathering of firewood, fetching of water, etc.). The ecofeminist perspective, particularly as presented in the work of Vandana Shiva (1988), criticizes Western science and development practice and links the oppression of women and the exploitation of the environment (Collins 1991). Ecofeminism recently has been criticized by Bina Agarwal (1991) and others for romanticizing the relationship of women with nature and for placing undue demands on women as a result of this perceived "special relationship." Leach presents evidence from Sierra Leone which delineates clearly the gendered power relations inherent in forest utilization by Mende women and men. She suggests that in the final analysis women's relationship with the environment is based both on negotiations with men and on a juggling of responsibilities and opportunities to take best advantage of those natural resources that are available to them.

In a similar vein, women's attitudes to natural resource management are not always more conservationist than those of men. Chimedza in Zimbabwe (1993) and Coulibally in Burkina Faso (1993), both argue that African women sometimes engage in environmentally destructive behavior, even when they are aware of the adverse consequences of their actions. For example, Shona women stripped branches from trees to collect small wildlife specimens (Chimedza 1993), while Burkina Faso women felled scarce trees to provide fuelwood for their families (Coulibally 1993). However in both cases, these environmentally destructive decisions were inspired by necessity rather than profit motivation. In this sense, women were forced to impose their own poverty on the environment.

This type of research, which has received support from some of the more progressive development agencies, approaches the relationship between women and natural resources from a new perspective, seeing women as active participants, decision-makers and negotiators in resource management. It begins with the recognition that women are not homogenous; that they have different interests, objectives and goals. Their relationships with men are not necessarily acrimonious but they *are* based on strategization and manipulation to achieve the best possible conditions for themselves and their children. However in understanding what constitutes a "best possible condition" we also must factor in non-economic considerations, particularly the underlying dynamics of male–female relationships and the cultural context in which they take place.

2 Science and Technology

"Science and technology" is commonly identified as a central component of economic development, but the term is often vaguely used and difficult to concretize. For development agencies working in a WID perspective, "science and technology" usually has meant appropriate technology, science education for women or, more recently, the entry of women into engineering or scientific occupations. For example, the *World Survey on the Role of Women in Development* (1986) stressed the importance of increased participation of women in science and technology training for ensuring that women's concerns are addressed in that realm. Similarly, UNESCO is currently expanding the provision of scholarships and opportunities for women to study and do research in science. These are important initiatives which should eventually counter the prevailing stereotypes about women and science, but at another level they serve to reaffirm existing conceptions of the science and technology/economics/development paradigm. However, they do not question the essentially male bias of the dominant science and development paradigm nor do they consider whether it adequately captures the interests, expertise and needs of women (or children).

Discussions of "science and technology" in the literature have often been divided into access and utilization questions, particularly with respect to rural women. Most commonly "science" and "technology" have been separated with the major emphasis on "technology." The literature on women, technology and development has tended to highlight the problems of technology utilization with relatively little consideration of women's roles as developers of technology. Again, women have generally been seen as passive recipients of technical ideas and knowledge gained from men. This has been reinforced by the tendency for girls to be actively prohibited or at least discouraged from pursuing technical training and skills. Even in cases where women have been the major users of technologies (e.g. in the case of household or food processing technologies), their role as developers of technology has been minimized.

Some analysts, like Filomina Chioma Steady, also have focused on women of the South as victims of technology transferred from the North in the context of global capital transactions (1990). This approach has given less attention to the options available to women themselves and tended to highlight the powerlessness of Third World women in the face of the dictates of international capitalism. This global perspective tends to minimize the issue of gender power relations at a local level, since men in developing countries can be seen as equally powerless with respect to international technology transfer.

Two common themes in the literature have been rural women's exclusion from access to technology and the displacement of women from traditional sources of employment as a result of the introduction of technologies that have been appropriated by men. Whitehead (1985) has noted that the most significant technological changes affecting women often are not aimed at them, and that these often have indirect and unforeseen consequences. She stresses the importance of analyzing the nature of women's work and the social context within which it is done. Boserup pointed out in the 1960s that as agricultural tasks became mechanized, they tended to become designated as exclusively "male work" (1970). Whitehead (1985) concurs, concluding that technological change in rural communities has generally reinforced existing class and gender inequalities.

Susan Bourque and Kay Warren (1990) have identified four distinctive approaches to gender and technology issues. The first encompasses what they call the "feminization of technology." Writers in this vein bring to the analysis of technology some of the same sentiments expressed by ecofeminists, differentiating between male and female values and arguing that technology must be redirected to serve more human purposes. A second perspective encompasses the appropriate technology movement with a focus on intermediate, locally-relevant technical solutions. This perspective has been criticized for encouraging policies that give women intermediate or so-called appropriate technology for work in the domestic realm while men are exposed to high tech. The third perspective sees women as workers in the global economy, their labor in both the productive and reproductive spheres being central to the maintenance of the current international system. It focuses also on the role of the state in legislating labor policy and analyses the nature of the interests and choices of policy-makers. A final perspective focuses on the ideologies associated with technical training and looks at the interaction between school and the workplace in reinforcing sex-role stereotypes. All of these approaches have generated significant findings. We believe some of the knowledge gained about the gendered nature of labor processes in terms of division of responsibilities, employment opportunities and the reconciliation of productive and reproductive work should be a starting point for analysis of gender and technology.

In trying to conceptualize a GAD approach to research and development projects in science and technology, one must begin by questioning the prevailing social structure of science and the implicit assumption that science and technology is neutral. As a number of feminists have pointed out (Haraway 1991; Harding 1986), science and technology is not only male-biased, it has been structured to minimize female participation except within the context of male-defined rules. Therefore, technical innovation is acceptable from an engineering or technical specialist but questionable from an unschooled user. Science and technology projects developed from a GAD perspective (drawing on feminist critiques of science as well) would challenge gender stereotypes by reinforcing women's confidence in their own observations as users of science and technology and enable them to work with technical specialists to improve technologies. Again, the issue of relative valuing of knowledge is of critical importance. While it would seem self-evident that the views and experiences of intended users of technologies should be a central source of information during the design of a new technology, this is almost never the case. The knowledge of women users is relegated to the category of the "unscientific or anecdotal" while the knowledge of technology designers (usually scientists or engineers) is considered "unbiased and relevant" (Haraway 1991). This strategy has led to the creation of thousands of unused technologies, and there is little evidence that more participatory approaches are being adopted. From a post-modern feminist perspective, this again provides clear indication of the necessity to listen to "other voices" and to assign appropriate value to different kinds of knowledge(s).

At a practical level, if it is clear that women have been displaced by the introduction of technologies, then what strategies can be developed to bring them back into the lost labor markets? If this is not practical then what new sources of employment can be developed? Will these new sources of employment be technologically-based? If so, how can we ensure that women have a knowledge of and a level of control over the technology? Most importantly, however, one must identify the sources of women's exclusion. Is it the result of systematic discrimination based on a set of assumptions about what constitutes "appropriate roles" for women? If this is the case, then women will suffer equally under a new program. It is important to attack the basis of these assumptions. Indeed, experience has shown that men often have appropriated work from women as it has become less labor-intensive and more technology-intensive. These are questions that can be effectively addressed by action-research undertaken from a GAD perspective, with a post-modern attention to difference, discourse and power. Women's own views and experiences must become the cornerstone of the policy and programatic options that emerge from such work.

In examining the approaches to sectoral development policy and practice, both in environment and in science and technology, it is clear that there is a tendency to begin from the safest and the easiest ground. Women and women's needs are inserted into ongoing projects and initiatives. Thus women are being given more information about environmentally sound farming practices; they are being given opportunities to study science; they are being taught to repair their own handpumps. These are important advances and should not be minimized. However one might question whether they will have long term benefits in the context of social, political and economic systems that are currently grappling with the limitations of a liberal development model that has been based on an assumption of slow but steady linear economic growth.

CONCLUSION

Projects undertaken from a gender perspective that assigns value to the experience and voices of all concerned actors are likely to have far-reaching implications. They will challenge the very structure of the societies in which they are based, and are likely to be politically sensitive and personally threatening to members of privileged elite groups. As has been noted, donor agencies tend to be inherently conservative and to operate in accordance with political interests. For the most part, programming decisions are based on supposedly neutral concepts such as "efficiency" and "effectiveness." Issues of power relationships between men and women and across social and ethnic groupings are rarely considered to be appropriate arenas for donor intervention. Nonetheless, there is some evidence that progressive individuals within some donor agencies are attempting to apply the ideology of GAD, with its transformative agenda, even while they use the language of WID (see chapter 12).

This chapter has suggested several possible reasons for the difficulty in operationalizing GAD projects but it may be worthwhile to focus further on what constitutes agreed-upon approaches in the field of development studies and practice and on the language used to justify and popularize different perspectives. As we have seen, development discourse is largely based on assumptions that have not changed substantially during the past thirty years and that never have been questioned very closely. Development practice has generally involved a heavy infusion of resources from outside with a predilection towards the "technological fix" (Stamp 1989). Development theorists and practitioners have learned little from past mistakes, nor have they fundamentally changed their way of thinking or their mode of operation. As a result, isolated knowledge in the form of case studies or academic papers generated in either the North or South has had relatively little impact on most development practice.

At the same time, we tend to minimize the recognition that the major actors in the development arena are both politically and economically motivated. In development planning and theorizing we seldom take into account the fact that donors seldom act exclusively from a sense of shared concern for the improvement of living conditions for people of the Third World but out of a desire to improve their own position. New power affiliations emerging out of development assistance have destroyed or eroded many traditional human relationships and values in the South.

Probably most significantly, as development theorists and practitioners we have frequently undervalued local experience and knowledge. This is being recognized as a failing only in the 1990s with the increased

219

attention to issues of "indigenous knowledge." While a romanticization of the past is undesirable, it has become increasingly clear that the wholesale rejection of community-based knowledge and practices advocated by development "experts" in the past, was not in the best interests of indigenous peoples nor indeed, compatible with the concept of sustainable development. As custodians of large segments of traditional knowledge, women suffered disproportionately from this rejection since their bases of authority were eroded or eliminated and for the most part were not replaced with new sources of authority (see chapters 3 and 10).

Although the Gender and Development approach has clear advantages over the WID approach in trying to integrate women's overall roles into an analysis of their participation in processes of social change, it has been less successful in integrating indigenous voices into its theoretical formulation (see chapter 2). However, the GAD perspective does offer some space for the correction of this fundamental wrong. Because it is not based on the notion of efficiency or linear progress, but instead tries to reflect the totality of women's experience and the nature of power relations with other actors in a given social context, it offers the possibility of broader interpretations. In this way too, the GAD perspective has sufficient flexibility to respond to the postmodernist call for the inclusion of "other voices." The embryonic movement towards the selective recognition, valuation and legitimization of traditional knowledge and experience provides some scope for optimism that development practice in the twenty-first century will be grounded in the realities of an ever more complex and interrelated world.

NOTES

1 An earlier version of this chapter was presented at the Association for Women in Development (AWID) meetings, Washington, DC, 20–24 November 1991. The author is indebted to Jane Parpart and Margaret Snyder for comments on the earlier version. The views expressed here are those of the author and do not necessarily reflect those of the International Development Research Centre (IDRC).
2 This is not meant to imply that bilateral and multilateral donors and nongovernmental organizations necessarily share the same approaches to or views of development programming. However, many NGOs, while perhaps taking a more humanistic approach to their work, have tended like larger donors to work within generally accepted definitions of development.
3 The desire to avoid "cultural interference" provided a rationale for avoiding the area of "Women in Development" by some agencies, notably the British Overseas Development Administration, until the 1980s (see Rathgeber 1988).

12

DECONSTRUCTING THE DEVELOPMENT "EXPERT"

Gender, development and the "vulnerable groups"

Jane L. Parpart[1]

INTRODUCTION

The development enterprise for the most part has been predicated on the assumption that certain peoples and societies are less developed than others, and that those who are more developed, i.e. more modern, have the expertise/knowledge to help the less developed (or developing) achieve modernity. The gap between the "developed" North and a "developing" South, and the assumption that development should follow a simple linear progression towards Western definitions of modernity, provided the rationale for the development business which has continued to expand since its inception in the 1940s. During this time, it has become increasingly professionalized. Universities have departments of development studies, and aspiring development experts can take courses in development policy, planning and practice. Some even receive diplomas attesting to their expert status. Indeed, much of development agencies' policy and planning is based on the premise that these experts, with their special knowledge of the modern, especially the technical world, are particularly well placed to solve the problems of the developing world.

The development expert has not been without critics, however. Concerns have been raised both within and outside mainstream development agencies about the relevance of certain projects as well as the dangers of dependency on foreign experts (Jaycox 1993; Nindi 1990). Some called for small-scale, appropriate technology and an end to large-scale, mechanized approaches to development (Schumacher 1973). Critics drawing on Marxist perspectives have turned development on its head, accusing Western development agencies, and their experts, of causing rather than alleviating underdevelopment in the South (Frank 1967). However, these critiques rarely questioned the assumption that Northern technical knowledge/expertise defined development and that the

221

transfer of that knowledge from the North to the South is at the core of the development process.

Of late, global economic restructuring, the emergence of new critical voices (many from the South and from women), and rising concern over the global environmental crisis have led some scholars and activists to rethink development assistance. Drawing on postmodern critiques of modernity, Western universalism, and dualist/binary thinking,[2] some scholars are taking the development debate in a new direction (Crush 1994; DuBois 1991; Edwards 1989; Escobar 1984–5, 1992; Ferguson 1990; Goetz 1991; Johnston 1991; Mathur 1989; Pieterse 1992). Recognizing the relationship between language and power, they have interrogated the language/discourse of development, particularly the (re)presentation of the South/Third World as the impoverished, backward "other" in need of salvation from the developed North/First World. This dualist construction, they point out, has reinforced the authority of Northern development agencies and specialists, whether mainstream or alternative, and provided the rationale for development policies and practices which are designed (whether consciously or not) to incorporate the South into a Northern-dominated world. This approach, they argue, is no longer appropriate in our increasingly complex and interrelated, if still unequal, world.

This chapter explores similar issues in the context of Gender and Development.[3] It draws on postmodernist feminist analysis, especially the question of knowledge production about women and the patriarchal character of Western (scientific) thought (Harding 1986; Hekman 1990). It analyzes the (re)presentation of Third World women by development experts, particularly the construction and dissemination of "knowledge" about Third World women, and the consequences of this knowledge production/discourse for both theory and practice. The chapter questions received wisdom about Women/Gender and Development, acknowledges the diversity of women's experiences in the South (and North) and calls for a more inclusive, multifaceted approach to development, development expertise and women's knowledge.

THE RISE OF THE "EXPERT"

The notion of expertise, with its accompanying certificates and institutions of learning, is so familiar today that it seems an inevitable part of life. But like so many things, expertise and the role of the expert has its own history. It is embedded in Enlightenment thought, which evolved in western Europe during the eighteenth century and which grew out of a growing belief in the ability of man (not woman) to apply rational, scientific analysis to the problems of life. The supernatural and other seemingly overwhelming forces were no longer seen as obstacles to (man)kind's efforts to understand and control nature (the environment). Rather, scientific knowledge and methods were increasingly seen as tools which would enable rational (male) individuals to bring progress and prosperity to humanity (Foucault 1986; Gutting 1989; Hekman 1990; Kant 1963).

During the nineteenth century, the Enlightenment project, as embodied in the industrial revolution and the rise of liberalism, led to the increased specialization of knowledge – witness the creation of new and separate disciplines within the academy and the creation of polytechnics in Germany and England. The growth of specialities led to a proliferation of specialists/experts. This specialized knowledge became increasingly associated with the rise of the new middle class, which in contrast to the nonspecialist "renaissance man" or humanist, acquired status and authority as the bearers of this new knowledge/expertise. As Michel Foucault has demonstrated, the emergence of new systems of control by the state, particularly in institutions of learning as well as prisons, asylums and other structures of state control, offered status, employment and authority for the professional experts in the new middle class (Foucault 1973; 1980). Indeed, professional knowledge/expertise became a crucial marker for identifying and maintaining the position of this class in the capitalist nation states of Europe and North America.

Control over expert knowledge became increasingly important and institutions to ensure control over the acquisition and use of this knowledge were developed and refined over the years. The legitimacy of the professional classes depended on widespread belief in their ability to define and transmit the scientific knowledge/truth needed by the modern world. Certificates testifying to the acquisition of "appropriate" knowledge provided a means of controlling access to this class and acknowledging and empowering those who had been given the authority to use this knowledge to discipline and regulate society. The gatekeepers in these institutions thus became the guardians of regimes of truth identified with modern societies and consequently played a pivotal role in the creation of Western modernity (Foucault 1980: 131–3). Western scientific knowledge was presented as universally valid and consequently applicable to all, but not everyone qualified as an expert. Increasingly, only the "properly" initiated could claim this title, and it is these "experts" who came to play a pivotal role in the process of collecting, controlling and transferring scientific knowledge between North and South.

DEVELOPMENT AND THE DEVELOPMENT EXPERT

As Europe and North America came to dominate the world, so too their assumptions about knowledge and development dominated relations with the South. The language changed over time, but the presumption of Western technical and moral superiority never flagged. During the eighteenth and nineteenth centuries, development was equated with "civilization," which was measured by the adoption of Western institutions and culture, especially Christianity. As such it was a highly gendered concept in which the Victorian woman was constructed as the benchmark of "civilization" as opposed to the *zenana* representation of non-Western women (Mohanty 1991a; chapter 1). But "civilization" was also affected by the assumption that Western technical knowledge, particularly scientific thought, was the highpoint of human endeavor and therefore the essence of "civilization" (Comte 1975).[4] Indeed, Western technical dominance was seen as a sign that the West, and all its institutions, were the litmus test of "civilization" and that in fact the two were interchangeable.

As Western society became increasingly secular and technical, the notion of Christian "civilization" was replaced by a belief in modernity, particularly economic and political development. The adoption of Christianity no longer defined the goal of human life, but the institutions and culture of the West, with its belief in the inexorable character of modernity/Western technical progress, spawned a science of economic development which measured all societies by the degree to which they had achieved the technical knowledge of Europe and North America.[5] In the early twentieth century, Western scholars increasingly compared the developed, modern economies with their less fortunate, undeveloped neighbours in the colonial empires of the world. The colonial world (later to become the Third World) became identified as the underdeveloped "other," in need of salvation by Western science (Johnston 1991; Mudimbe 1988; Pieterse 1991; Stein and Stein 1970).

Gradually, the "science" of development economics evolved, and with it the assertion that economic "laws" could solve developmental problems. As the world, including the South, was increasingly seen as a collection of economies rather than societies, development economics gained increasing prominence as the "science" of economic progress. Those nations identified/constructed as (economically) less developed were seen as particularly in need of this expertise. This discipline promised that with judicious effort and the assistance of developmental experts, less developed economies could eventually be brought up to modern standards (Rostow 1960). Only then, and only with Western assistance, would they become full partners in the modern world, ready to participate in the adventure of progressive modern development already well under way in Europe and North America. Of course, the discourse and institutions of modernity had already been set by the West. Third World peoples had merely to adopt these, not to introduce

224

"traditional," i.e. old fashioned, concerns (Johnston 1991; Mathur 1989; Watts 1993).

The early experiments in Third World development which began in the 1920s and 1930s, and increased dramatically after World War II, were deeply embedded in Enlightenment thinking, with its belief in trained, qualified scientific expertise. They were generally large technically oriented projects, often poorly thought out and executed (Grischow 1993).[6] These projects were staffed by development experts whose training soon became institutionalized in universities, which established programs in development studies, development economics and public administration, along with many other developmental technical subjects. University certificates testified to development experts' legitimate status and provided legitimation for consultancies and employment in Northern development agencies and programs. These experts became, and continue to be, essential to the development enterprise, as development policies and programs are largely predicated on the assumption that developmental problems can be reduced to technical, i.e. "solvable," problems which involve the transfer of Western technical expertise to the developing world (Escobar 1992; Johnston 1991; Manzo 1991; Nandy 1989; Nindi 1990).

The development business, with its ever larger bureaucracies, has been largely built on this assumption, and the transfer of expertise from developed to developing countries has been the underlying rationale and practice of development agencies and practitioners. Internal criticisms, such as those by the World Bank Vice President, Africa Region, Edward Jaycox, have focused on particular problems, such as the deleterious impact on professionalism in the Third World caused by "the tendency to use expatriate resident technical assistance to solve all kinds of problems." He calls for the creation of a "demand for professionalism in Africa," but never questions the authority of the North over the definition and transmission of that professional/expert knowledge (Jaycox 1993). Like most internal critics, he remains locked within the basic assumptions of his institutional context.

The dramatic failures of some large mainstream development projects in the 1960s and 1970s spurred criticisms from outside the development establishment as well. In 1976, for example, a Tanzanian scholar complained about the endless, underutilized and often faulty development plans in Tanzania (Lwehabura, cited in Armstrong 1987).[7] Scholars adopting a neo-Marxist dependency approach, particularly in Latin America, accused Western development specialists of being agents of capitalism, determined to perpetuate Third World underdevelopment (Frank 1967; Kay 1989). More recently, post-Marxists have declared an impasse in development theory and called for new, more nuanced approaches to development (Booth 1985; Corbridge 1989; Schuurman

225

1993a). However, none of these critiques challenge the underdeveloped status assigned to the Third World, or the belief that modernity/development requires technical assistance from the North. Moreover, the superiority of Northern expertise is widely accepted in the South as well (see chapter 10).

However, the sanguine belief in the North's privileged position has become more difficult to sustain in the face of global restructuring, widespread environmental crises and the emergence of new voices demanding to be heard. Old certainties have been called into question, especially the assertion that only Northern expertise can solve the world's developmental problems. Of late, scholars drawing on postmodernist thinking have sought a new perspective on development. Acknowledging the connection between power and language, Ferguson (1990), Escobar (1984–5; 1992), DuBois (1991), Wood (1985), and others[8] have focused on the discourse of development. They have interrogated the language of development policies and planners, particularly the way they (re)present their Third World clients, the solutions proposed for those clients and the role of development specialists in this process. Ferguson (1990), for example, discovered that a CIDA/World Bank project in Lesotho (re-)presented Lesotho as an isolated country in need of Northern expertise, ignoring years of contact between Lesotho and South Africa, the cultural norms that would be breached by the proposed cattle sales scheme, and the political consequences of the project (see chapter 1). This example, and many others,[9] raise questions about knowledge production and dissemination by development specialists. They remind us that control over discourse does not simply operate at the level of language. It reinforces and legitimates policies and practices that have strengthened Southern ties with the North and consequently Northern influence in the region. It has also silenced dissenting voices and buttressed the vision of a homogenous South best understood and assisted by Northern or Northern-trained experts.[10] While recognizing that positive, often unintended, consequences do sometimes emerge from development projects, these scholars question whether self-reliant, locally situated development can occur as long as development remains embedded in modernist thinking. They call for new ways of thinking about development theory and practice (Escobar 1992; Esteva 1987; Ferguson 1990).

The question here is whether these same issues, and perhaps others raised by postmodernist feminists, have anything to tell us about the Gender and Development enterprise. We are particularly concerned with how the issue of representation, with its roots in patriarchal Western thought, has influenced development policy and practice for women.

DEVELOPMENT DISCOURSE, WID AND THE "THIRD WORLD WOMAN"

During the first few development decades (the 1950s and 1960s), development theory and practice paid little attention to Third World women. As we have seen, development was regarded as a technical problem, one that required male expertise from the North and male cooperation in the South. Women and children were regarded as potential beneficiaries (after all, they walked on roads too), but not as potential agents for change. This approach was based on Western stereotypes which represented women primarily as mothers and wives rather than as economic actors who might produce goods (other than people) and even sell those goods in the market. Hence women entered into discussions of population control, child health, nutrition and little else (Buvinic 1989; Mueller 1987). Drawing on a long history of colonial discourse which represented Third World women as particularly backward and primitive, development planners continued and even extended the representation of Third World women as the primitive "other," mired in tradition and opposed to modernity. This, of course, legitimated leaving women out of the policies and plans of development experts (see chapter 1).

As a result, early development projects for the most part ignored women's input, despite their pivotal role in the economic life of most Southern countries. Not surprisingly, in 1970 Ester Boserup reported that most development projects not only disregarded women, they frequently hurt them. Indeed, she pointed out that many projects had failed *because* they ignored women's work and advice. Boserup's revelations initially had little effect on development practice. In the 1970s, for example, a large NORAD-funded settlement scheme in a matrilineal region of Zambia foundered at the implementation stage because it had disregarded the views of local women (NORAD 1980). Indeed, in 1979 Barbara Rogers reported many more such failures.

However, the need to focus squarely on women, and to provide special development programs for them gradually gained credence among the more progressive development agencies, especially in the Nordic countries. The sub-field of Women in Development (WID) gained increasing legitimacy in development circles, and development specialists concerned with women began to establish footholds in mainstream development agencies (Kardam 1989; Tinker 1990; chapter 11).[11]

Despite various changes over time,[12] WID policy and practice has remained consistently grounded in the assumption that women need to be integrated into the development or modernization process. The state, both local and foreign, is seen as an ally ready to assist this process. This benign perspective towards the state by WID practitioners has led to policies and projects that rarely question the priorities of male-dominated host governments.[13] Basic assumptions about women and men, the accepted sexual division of labor, and traditions that bind women into subordinate positions are seen as sacred areas that must be left alone. They are reified as culture, and therefore placed outside the development mandate (Goetz 1991; Mueller 1987; Stamp 1989). This is, of course, a position that is encouraged and applauded by many male leaders in the South, who are perplexed and annoyed by Western development agencies' persistent talk about women's development (World Bank 1989c: x). Small income-generating projects, women's bureaus and units, bursaries and gender-sensitization workshops are fine as long as long as they are funded largely by development agencies and have no real power (Overholt et al. 1985; Parpart and Staudt 1989; chapter 11).[14]

WID discourse on Third World women and their problems has reinforced these assumptions. A consistent theme runs through their document, whether academic or policy-oriented. Third World women emerge as "backward, premodern beings," with no agenda of their own, tied to traditional ways of thinking and acting. As a World Bank study put it:

> Culture and tradition vary but often confine women and girls inside the family or close to home . . . . As a result, women's productivity is frequently depressed well below potential levels – and this carries a cost in economic efficiency. Women are, in a sense, wasted . . . women feel reluctant to seek help for themselves or their children. . . . In some societies where women are not encouraged to *think* for themselves, authority figures have helped persuade women to seek health or family planning services, continue breastfeeding, and so on.
>
> (World Bank 1989c: iii)

This construction/representation of Third World women has been reinforced by the current economic crisis, which has highlighted the characterization of women in the South as the *vulnerable* "other," victimized by the retrogressive traditions and economic ineptitude of Third World economies. The Commonwealth Expert Group on Women and Structural Adjustment publicized the phrase the "vulnerable groups" to describe the dire consequences of SAP policies for women and other disadvantaged groups in the South (Commonwealth Export Group 1989). This evocative phrase became a rallying cry for development experts wishing to challenge World Bank policies, particularly from UNESCO (Cornia et al. 1987). While an effective weapon against the

Bank, this language has further entrenched the image of the helpless premodern, vulnerable Third World woman.

This discourse legitimates the need for development aid, especially technical aid, by constructing Third World women's problems as technical problems requiring a technical (usually Northern) answer. Statistics on women chronicle their many problems, emphasizing their subordinate status, their high mortality, lack of education, poverty and powerlessness. Variations between rich and poor are submerged in a sea of negative numbers (Joekes 1987; Mueller 1987).[15] Third World women are constructed as a homogenous group, desperately in need of Northern technical assistance. These assumptions underlay the Sustained Health Improvement through Expanded Livelihood Development (SHIELD) project in the Philippines, for example. The project adopted a narrow definition of medical knowledge, and consequently "discovered" widespread "medical" ignorance among the women in the communities being assisted. As a result, the project saw its mandate as the transmission of "modern" medical knowledge to "ignorant" women. This was seen as a technical problem, which required expert development assistance. The possibility that local women might have useful, and even crucial, indigenous medical knowledge, was never considered (St. Hilaire 1992).

WID discourse not only established the need for technical assistance from the North or Northern experts, it also reinforced the authority and power of the WID development expert, who became the "saviour" of the vulnerable groups. In the SHIELD case, medical knowledge was represented as residing solely in foreign or foreign-trained experts, whose job was to transmit that knowledge. The glorification of the expert and expert qualifications legitimates the authoritative role of the WID expert, particularly the hierarchical nature of skill transmission, and the belief that modernity and "modern" knowledge can only be acquired through Northern assistance.

WID discourse has also encouraged development practitioners (and scholars) to undervalue knowledge that comes from living in poverty, from working out solutions to daily life in specific, often difficult, locales, and from cultural traditions that have provided basic but adequate survival patterns for hundreds of years. This knowledge has rarely been regarded as a source of enlightenment; rather it has for the most part been seen as a barrier to development. Indeed, the same World Bank study argues:

> Women are bound by tradition and gender based difficulties. . . . To improve women's nutritional status, women themselves must be convinced of the need. In many circumstances, creating demand is as important as providing services. . . . Women's lack of self-con-

fidence is a major impediment to the success of maternal and child health programs. It often shows up as "silence" or extreme denial of self and dependence on external authorities for direction.

(World Bank 1989c: iii, 35)

The possibility that Third World women know how to act in their own interest has been largely ignored. This is particularly apparent in discussions about the rural and urban poor (see chapter 2), but the tendency to homogenize Third World women's condition, and to see all Third World women as poor and ignorant, has also undermined belief in the capacity of Southern development expertise, especially among women (de Boef *et al.* 1993; Wiltshire 1988).[16]

What have been the consequences of these policy prescriptions and assumptions for the practice of development? This is, of course, a complex subject. Policy does not necessarily translate into practice, and unintended consequences abound. As we have seen, WID discourse claims specialized knowledge about both Third World women's problems and their solutions. However, these claims are not always heard. In large technical projects, the opinions and suggestions of WID specialists have often been drowned out by (usually male) hostile technocrats from the North and South (Braxton 1991). Large projects pay lip service to gender issues, but often do little about them. Agricultural training continues to be mostly aimed at men, often with disastrous results. In Burkina Faso, for example, in the 1980s a village livestock project provided training to men when women were responsible for small livestock production. Predictably, the project failed (Carloni 1987). Extension workers continue to be predominantly men, even though in many parts of the world, women do most of the agricultural work.[17]

But more women-centered projects have not always done much better. The Commission of the European Community evaluated nine projects on women in rural development which it had funded and discovered that five of the projects reported little or no participation by the women supposedly involved (CEC 1990). In Swaziland, programs to encourage micro and small-scale enterprises developed by the Dutch in the 1970s provided skills training with little relevance to Swaziland's industrial policy. Indeed, they increased competition over handicraft products and left the women no better off than before (Gosses n.d.). Mainstream projects have often done more to increase government control over women than to improve their lives. The SHIELD project drew women more tightly into community and government structures while weakening existing women's organizations. The complex, technical language and structures of most WID projects often marginalize local women, but they also reinforce the authority of the technical (often foreign) experts in the project. In the SHIELD project this led to cumbersome, time-consuming activities with little relevance for local women, male control over much of the borrowed funding and a studied refusal by the project leaders to listen to women's opinions in the matter (St. Hilaire 1992). Of course, as Sylvester points out, women sometimes appropriate offical knowledge for their own ends as well (see chapter 10).

Of late, some development agencies have deliberately involved local women in their projects. This is crucial and needs to be encouraged. However, local personnel are often given peripheral responsibilities with little autonomous authority (Opole 1993). The possibility that local experts may also be outsiders among the poor, and that they may have adopted Western attitudes toward their less educated sisters, is rarely considered. Nor do most project managers worry about the agendas of local elites and local technocrats, who, as St. Hilaire points out, may have their own agendas for the poor, agendas which often conflict with the goals of those being "helped" (Mueller 1987; St. Hilaire 1992).

Many of the mainstream development agencies have sought cooperation with non-governmental organizations (NGOs) in order to pursue more grassroots development efforts. These alliances may be with large international NGOs, such as Oxfam, or with more successful local NGOs. These partnerships have often provided services which large-scale development projects cannot supply, or they may support activities which are usually not the purview of mainstream development programs. For example, the Danish development agency, DANIDA, has funded WLSA (Women and the Law in Southern Africa), which has carried out path-breaking research and participatory action on women's legal issues. However, many of these partnerships exhibit the same insensitivities as larger mainstream projects. For example, a large watershed project in India, run by the Swiss Development agency and a local NGO, Myrada, arranged meetings with farmers in the evenings, with no discussion of whether this was a suitable time for women. Moreover, the project leaders lumped women in with other vulnerable groups such as "tribals and landless people," and offered them special non-formal education "to assist them to realize their own potential and break out of the existing situation." The possibility that women (or the other "vulnerable" groups) might have anything to offer the project was never mentioned (Mascarenhas 1993).

Of course, some WID projects have managed to provide useful and even important opportunities. Individual development practitioners, who have a "feel" for the social milieu in which a project is being carried out, are sometimes able to direct projects in ways that truly serve women's needs (chapter 11; deBoef et al. 1993). Project leaders have learned from mistakes. The NORAD settlement scheme in Zambia, for example, eventually took women's interests into account, sought to register women as landowners to the same extent as men and reserved half the training places for the women (NORAD 1980).

However, unintended consequences of development projects are frequently more interesting than intended outputs. Ferguson (1990) discovered that women had used the Lesotho project's cattle sales program to obtain resources usually unavailable to them (as cattle are traditionally a male preserve). This was entirely unexpected. As Wood (1985: 368) points out, resistance to a project may be the most fruitful reaction possible if the project is designed to disorganize one's class and subordinate one to the state. For while many projects have increased official control over marginal peoples and thus buttressed the authority of the indigenous elites, development recipients have sometimes used skills and other opportunities provided by a project to challenge these heirarchies. Skills training has been used to challenge the government or to organize in unintended ways. Sylvester (chapter 10) discovered that her informants increased their own authority in cooperatives by associating their knowledge with the official knowledge of technical experts. Credit and employment opportunities have enabled women to challenge the sexual division of labor in the household and even gender hierarchies in the community (Chimedza 1993; Esteva 1987; Kandiyoti 1990). For example, women have used credit from the Grameen Bank to obtain economic independence. Women in the CEC-funded rural projects told the evaluators that, above all, they wanted their own incomes so they wouldn't need to depend on remittances from often unreliable, migrant husbands (CEC 1990). WID projects almost never mention political outputs, as they are rarely designed to foster political action.[18] But they often do anyway. One has only to look at the political movements emerging in Latin America and the environmental movements in Africa and Asia (Mies and Shiva 1993; Shiva 1988). In fact, although the bane of development planners and evaluators, unintended consequences are often the most meaningful part of a development project.

ALTERNATIVE DEVELOPMENT AND THE "THIRD WORLD WOMAN"

WID is not the only approach to women's development issues. Influenced by both critiques of mainstream development and radical feminists' call for separate development, in the 1970s some scholars and activists became convinced that self-reliant development could not happen through reform from within established power structures. These ideas influenced the policies and practices of some development agencies, particularly the NGOs, which pride themselves on their grassroots approach to development and their greater knowledge of and sensitivity to Third World peoples.[19]

Indeed the discourse of development agencies committed to this approach are replete with assertions about the need to avoid governmental interventions as much as possible, to keep development projects small, to respect the knowledge of the local peoples and to adopt a participatory approach to development in order to avoid domination by development experts and the smothering of the grassroots point of view. In the case of women, this led to arguments for small-scale women-only projects in order to avoid male domination, and to participatory approaches so grassroots women had a say in the construction and carrying out of the development project. This approach became known as Women and Development (WAD) and has had considerable influence in what is sometimes known as alternative development circles (Parpart 1993; Rathgeber 1990).

While apparently quite different from the WID approach, the representation of Third World women by development experts working within this approach often betrays neo-colonial leanings. For example, the introduction to the Oxfam *Field Directors' Handbook* emphasizes the need to respect and listen to the opinions and needs of Third World women, and to foster development "in such a way that the poorest are enabled to take charge of their own lives" (Pratt and Boyden 1988: 12–13). Yet the section on women, sounding very much like WID, warns that "women will often, by virtue of their role in society, accept without question prevailing norms and will not challenge the injustice they experience." It goes on to characterize women in India as "brought up to play what has been called a 'life-long role of subservience and self-effacement'" (*ibid.*: 43). The *Handbook* focuses on the impoverishment of female-headed households, the decline in women's health and its adverse impact on children, lower literacy rates and patriarchal domination. While this data is correct for many women, the implication that it applies to *all or even most* Third World women, reinforces neo-colonial and WID stereotypes of the Third World woman. An IDRC report (Gahlot 1991) on an NGO dedicated to helping squatters in Bombay (SPARC – Society for the Promotion of Area Resource Centers) illustrates this tendency as well. When discussing SPARC's successes, the leaders emphasize their inputs, rarely mentioning the commitment, knowledge and street sense of the pavement dwellers themselves. This representation suggests an unwillingness to recognize and encourage self-reliant development among the pavement dwellers.

The tendency to represent Third World women as hapless and vulnerable has affected the practice of alternative development agencies. Despite protestations to the contrary, many of these agencies and their experts (sometimes called cooperants to sound more egalitarian) fall into some of the same patterns of development practice as those in the more mainstream agencies (see chapter 2).[20] Some of this behavior is no doubt encouraged by the need to create technically feasible projects, which can be administered, budgeted and evaluated on the basis of measurable inputs and outputs. However, this behavior is also grounded in the representation of Third World women as vulnerable, helpless victims. Indeed, the cooperant can be just as much the development expert, who "knows" the answers to people's problems, as the mainstream WID expert. A foreign-funded NGO program delivering primary health care (PHC) in Burkina Faso in the 1980s, for example, focused largely on imparting scientific medical knowledge to villagers, ignoring local medical knowledge and health practices, as well as the political and social structures of the villages in question. As a result, dependence on PHC experts increased, while the project's public health training had virtually no impact. The women involved simply reverted to traditional practices when left on their own. Indeed the project did more to integrate the villages into the government bureaucracies than to improve women's self-reliance or alter entrenched patriarchal structures (Maclure forthcoming).

While NGOs have often provided crucial services to the poor, certain problems have surfaced repeatedly. Beckley's study of twelve NGOs in Sierra Leone illustrates a few of them. She discovered that the NGOs played a "vital role in meeting the basic needs of underprivileged women, especially in the rural areas." However, the programs were often short-lived. They often ignored the "central issue of gender and the interaction between gender and development" (Beckley 1989: 51–2). Moreover, they often supplemented government services, thus propping up unsavoury regimes and encouraging the status quo. They also too often fostered unrealistic expectations, a dependence on NGO assistance and disillusionment when that assistance was not forthcoming (*ibid.*: 52–5).

Some community activists and scholars are beginning to call for serious rethinking of alternative development practices. They are usually people who have been involved in projects long enough to respect the knowledge of their "clients" and to learn from their own mistakes. The Power Stations project in Grahamstown, South Africa, is a case in point. Created by local, generally well educated, liberal whites, to foster skills and production in local handicrafts, the project has been an important source of employment, skill transmission and pride in a community with a 70 percent unemployment rate. It grew rapidly as its success attracted attention and funding. Yet, when the white project leaders withdrew, political divisions and dependence on this expertise led to a dramatic decline in productivity and the eventual closing of the project. It has recently reopened with a much smaller staff, a clear realization that transferal of skills does not in itself solve developmental problems, and the determination that skills will be taught in ways that foster independence and self-sufficiency. In order to do this, the project leaders have involved local participants in planning and implementing skills training so the process could be designed to fit local cultural, social and political patterns. This has required a rethinking of the process of skills transfer and the recognition that knowledge in and of itself cannot ensure development (Ann Walker, project leader, Grahamstown, South Africa, 16 August 1993). This experience suggests that those experts who have learned to respect the knowledge of their clients are usually people who have spent long enough in a project to learn from their own mistakes. It also suggests the need to reconceptualize our understanding of Third World women's development expertise and its crucial role in skill transmission, no matter how much this transfer appears to be simply a technical matter.

This reconceptualization/reconstruction of the development subject and the character of development knowledge is not easily achieved. It requires time and many painful experiences. It also requires humility and a persistent refusal to foster recipient dependence (Bill Davies, Director, Community Development Studies, Rhodes University, Grahamstown, 22 August 1993). While exceptions occur, and many NGOs are introducing more sensitive, people-oriented development projects, the belief that alternative development practices can easily transcend many of the hegemonic beliefs and practices of larger, more mainstream development agencies is clearly untenable.

GAD, REPRESENTATION AND THE DEVELOPMENT OF WOMEN

Has the Gender and Development (GAD) approach overcome these limitations? As we have seen, GAD is situated squarely in both socialist feminist scholarship and the writings of Third World feminists, particularly the DAWN group. This approach seeks fundamental explanations for women's subordination, both at the level of broad political and economic forces and at the level of ideology, particularly gender ideology. It acknowledges the need to understand gender relations on the ground, and to investigate the specific ways gender ideology and relations contribute to women's subordination and the sexual division of labor and power. Third World feminists' influence on the GAD approach no doubt accounts for its emphasis on poverty and global inequalities (Goetz 1991; Kabeer 1991; Sen and Grown 1987).

GAD calls for both short-term and long-term approaches to women's development, and to a gender-sensitive rather than a woman-only approach. The short-term goals of GAD specialists are often cast in much the same language as WID, i.e. they involve education, credit, improvements in the legal system, etc. The long-term goals include ways to empower women through collective action, to encourage women to challenge gender ideologies and institutions that subordinate women. The challenge to established structures and norms has been largely unacceptable, particularly to international and governmental development agencies which regard sovereign rights as sacred and reject social transformation as a developmental issue (Kabeer 1991; Moser 1993; Young *et al.* 1981; chapter 11).

The GAD approach is a breakthrough in many ways, but it too often falls into modernist stereotypes. Third World women are frequently represented as the impoverished, vulnerable "other." For example, the key DAWN document by Sen and Grown (1987) portrays women in the South as helpless victims, who need to be saved from poverty and backwardness. The capitalist system is blamed for both Third World poverty and women's subordination and poverty. According to Sen and Grown:

> For women this vulnerability [i.e. being poor in the South] is further reinforced by systems of male domination that, on the one hand, deny or limit their access to economic resources and political participation, and on the other hand, impose sexual divisions of labor that allocate to them the most onerous, labor-intensive, poorly rewarded tasks inside and outside the home, as well as the longest hours of work.
>
> (1987: 26)

The impact of this construction/language is never considered, nor are the solutions to this condition posited outside the realm of development/ modernization. Expert knowledge is still regarded as the "solution" to women's developmental problems. Moreover, the adoption of a gendered approach is seen as requiring expertise held largely in the North or by indigenous experts trained in the North (Kabeer 1991; Sen and Grown 1987; chapter 2).

At the level of planning and practice, this modernist discourse has reinforced the hierarchization between formal development credentials and the expertise of indigenous peoples. For example, the mechanisms for integrating gender into planning and policy are often as complex and difficult as most mainstream planning. Moser describes planning cycles, data collection, monitoring and feedback procedures that would exclude all but the most formally educated women in the South (and North) (Moser 1993: 84–5). The formalities of planning thus reproduce, intentionally or not, hierarchies within the development enterprise. The assumption that modernization requires such planning (and expertise) inhibits new ways of imagining development, at the level both of theory and practice. It restricts much GAD analysis to modernist definitions of development with their emphasis on Northern (or Northern-trained) technical expertise and the universal applicability of Northern models to development problems in the South and North.

THE EMPOWERMENT APPROACH[21]

Of late, some scholars and activists in the South (and a few in the North) have begun to question the discourse and practice of development, whether mainstream or alternative. Reflecting changes in the global political economy and growing concern for global environmental crises, new voices, very often from the South and from women, are entering the development debate. The North/South divide, Northern authority over the "modern,"[22] and Northern claims to "know" Third World women are being called into question. While drawing on and contributing to various international debates (see chapter 1), these scholars and activists above all argue for a new definition of development, one that is grounded in the experience(s) and knowledge(s) of women in the South. Bina Agarwal calls for:

> an alternative transformational approach to development [which] would . . . concern both how gender relations and relations between people and the non-human world are conceptualised, and how they are concretised in terms of the distribution of property, power and knowledge.

> (Agarwal 1991: 58)

Virginia Vargas emphasizes the need to acknowledge diversity among feminist scholars and activists in Latin America. She points to new developments, to new voices that must be incorporated in the development process:

> There will be many more women all over the continent expressing multicultural perceptions and revealing a variety of complex differences. New participants will demand their own space. We must work out new ways of meeting, of keeping that plurality from washing us away, of creating institutions that can combine pluralism with individual choice and intiative.

> (1992: 212)

Heyzer and Wee criticize the blind adoption of the growth model of development in Southeast Asia, and call for new ways of thinking about development and the environment that are grounded in women's daily lives (Heyzer and Wee forthcoming). While the themes vary, the general message is clear – development theory and planning for women must exhibit greater sensitivity to difference and an awareness of the multiple oppressions – particularly race, class, ethnicity and gender – which define women's lives in the South (and minority women in the North).

Only then will the complex, multilayered lives of Third World women (and men) be understood and their problems addressed (Barriteau 1992; Heyzer and Wee forthcoming; James and Busia 1993; Mbilinyi 1989; Vargas 1992).

These concerns have been reflected in practice as well. Women have joined the new social movements and used them as a platform for their ideas and demands (Escobar 1992; Heyzer et al. 1995). The study and practice of women's political action has begun to focus on difference, multiple identities and discourse as well (Nzomo 1992; Vargas 1992). Postcolonial critics in the South are interrogating the language of postcolonial and neo-colonial writings and finding new ways to envision resistance (Bhabha 1994; Rajan 1993). Environmental activists, such as Vandana Shiva, are speaking of the need to decolonize Northern assumptions about the environment and development, particularly the notion that sustainable development can solve environmental problems without necessitating a reevaluation of development itself. Shiva calls for a new approach to the environment, one that puts women first and draws on the environmental knowledge of poor women in the South (Mies and Shiva 1993; Shiva 1988). This theme, along with postmodernist feminist critiques of science and knowledge (Harding 1986; 1992; Hekman 1990), has inspired attacks on Western science and the growth model of economic development (Heyzer and Wee forthcoming; Mies and Shiva 1993; *Review of Women's Studies* 1991-2). Many of these concerns are being expressed in the North as well, and they are spurring the creation of new networks, such as WIDE, that acknowledge differences without abandoning the commitment to feminist agendas (Connelly et al. forthcoming; Braidotti et al. 1994; Beneria and Feldman 1992; Harcourt 1994; see Introduction/Conclusion).

CONCLUSION

The debates about the development enterprise are clearly ongoing and fluid. However, they suggest a widespread scepticism in the South (and among some people in the North) about Northern models and prescriptions for Third World women (and men). We are at a crossroads. New voices from the South are entering old debates and changing the terms of our understanding. New regions of prosperity are providing the support and confidence to question Northern control over the definition of modernity, although new definitions are sometimes redefining development in patriarchal and modernist ways (Ong 1993). At the same time, the continuing crisis in parts of the South, particularly Africa, is undermining the credibility of structural adjustment programs. Even mainstream development agencies are beginning to recognize the need for human capacity building in the South and the bankruptcy of many supposedly "failsafe" development solutions. Increasingly, mainstream and alternative development advocates are calling for greater participation by development partners (Jaycox 1993; Moser 1993; UNDP 1993). Indeed, indigenous knowledge was a declared a key tenet of the UNCED Agenda 21 Meeting in Rio in 1992, and the search for local knowledge has become a staple goal of many mainstream development agencies. However, the preeminence of Western science and the modernist framework is rarely questioned, and relations between indigenous knowledge brokers and Western (or Western-trained) development specialists for the most part remains problematic (de Boef *et al.* 1993: xii; Opole 1993).

However, some scholars and activists recognize the need to move beyond modernization, to adopt an approach to development that pays more attention to the different voices/knowledge(s) in the South, and that designs policies and practices based on the concrete, spatial and environmental contexts in which people live (Alatas 1993; Esteva 1987; Hobart 1993). An indigenous vegetable project in Kenya is trying to wean scientists away from a preoccupation with measuring outcomes and products, and to encourage openness to less quantifiable but often more appropriate local knowledge(s) (Opole 1993: 159). Nandy emphasizes the need to pay attention to unorthodox knowledge, such as that of the Shaman who is outside "civilized" Western culture and thus offers a basis for and symbol of resistance to the dominant politics of knowledge (1989). Blaney and Inayatullah advocate "a conversation of cultures . . . [which] depends on the capacity and willingness of the winners and losers to distance themselves from their position in the global hierarchy and bring themselves closer to the other" (1994: 24, 45). Escobar calls for a reassessment of basic needs that is grounded in the experiences of people's everyday experiences in ways that "bypass the rationality of development with its "basic needs" discourse" (1992: 46).

This more interactive, open and critical approach to development is certainly a step in the right direction. But translating these ideas into practice is a difficult and often neglected task. Participatory action research (PAR) offers some useful pointers. It recommends openness and the need to listen to others (Esteva 1987; Fals Borda and Rahman 1991), but it underestimates the difficulties inherent in such a task. For this, particularly in regard to Gender and Development, postmodern feminist thinking has much to offer. Truly listening to others entails moving outside your own conceptual frameworks, especially the binary thought structures and patriarchal character of most Western knowledge. It requires the recognition that differences, and different voices, cannot just be heard, that language is powerful and that subjectivity (voices) are constructed and embedded in the complex experiential and discursive environments of daily life. Overcoming these barriers is not easy. In Bali, for example, Mark Hobart discovered that "Knowledge is not objectified or commoditized . . . the stress is not on the text or its author as agent, but on the interpreter as agent in collaboration with those concerned." To "hear" such differences, he argues, we must learn to avoid prejudging what knowing is. We must "start with situated practices: what people did and what people said about it". Above all, we must acknowledge that "Knowing is not an exclusive prerogative of some superior knowing subject" (Hobart 1993:15-16). Thus he reminds us of the immense complexity involved in communication, of the power of language to construct knowledge in ways that inhibit understanding. Clearly, listening and communicating cannot be assumed; they must be problematized and worked on (Connelly *et al.* forthcoming; Goetz 1991).

Cooperation based on equal partnership between Northern (and some Southern) experts is also rare and difficult. Most partnerships between North and South have focused on transmitting information *from* the North *to* the South, or from Southern experts to the poor. Many have been "fraught with tensions and conflicts" and have failed to produce the expected benefits. Partnerships, like communication, need to be problematized. They require explicit understanding of expectations and assumptions, particularly the modernist hierarchy implicit in much development discourse and planning, along with detailed discussions of logistics and lots of good will. These are rare, and will continue to be so until the difficulties inherent in joint activities is acknowledged and dealt with (Sherrill Johnson, personal communication, 16 June 1994).

Above all, development planners and practitioners must abandon their belief in the invincibility of Northern-based technical expertise. This does not require the abandonment of technical assistance, nor the rejection of technology transfers. But it does require a fundamental rethinking of modernization, and development, an interrogation of scientific knowledge and its role in development, and in the case of women, a clear recognition of the patriarchal nature of much Western knowledge. We need to abandon the North/South divide, with its arbitrary and old-fashioned division of the world, without merely adopting other patriarchal knowledge structures. Emancipatory development will only occur when development theorists and practitioners adopt a more inclusive approach to knowledge/expertise, a readiness and ability to "hear" different voices/experiences, and the humility to recognize that established discourses and practices of development have often done more harm than good. Then, and only then, will development expertise take its rightful, more modest place, in the search for a better, more equitable world.

NOTES

1 This chapter has benefited enormously from comments by Marianne Marchand, as well as from discussions with my colleagues Pat Connelly, Tania Li and Martha MacDonald.

2   Western thought is often captured in dualist or binary pairs, where one side of the pair (such as man/woman, nature/nurture) is believed superior to and the opposite of the other. To deconstruct these binary systems one looks for meanings embedded in the oppositional pairings (Culler 1982).

3   By women and development I am referring to all the various approaches to women's development issues, not to the specific approach which is known as Women and Development (WAD) to be discussed later in the chapter.

4   Auguste Comte identified civilization with a hierarchy of knowledge, wherein positivist, scientific knowledge replaced theological and metaphysical approaches to knowledge. Comte believed the natural sciences were the pinnacle of human intellectual endeavors. This assumption enabled Western scholars to silence, denigrate, ignore and subjugate knowledge which was not classified as scientific. This includes most of what we call "indigenous knowledge."

5   It should come as no surprise that the "science" of economics is the most masculine of the social sciences (Manzo 1991; Marchand 1994).

6   For example, in the 1950s the British government established the Damongo Groundnut Project in Ghana. Ignoring local knowledge, and misrepresenting local conditions, the project administrators blithely assumed Western technical expertise could overcome all obstacles. Not surprisingly, the project was a complete disaster (Grishchow 1993; see also Roe 1991).

7   The World Bank loaned $60 million to Tanzania in the 1970s to create the gigantic Mufindi Pulp and Paper Mill, now regarded as the worst "white elephant" industrial project in Tanzania's history (Nindi 1990; 55). Similar complaints surfaced in all the different regions of the South (Escobar 1992; Ferguson 1990; Heyzer and Wee forthcoming).

8   Scholarship on development issues has increasingly drawn on postmodernist thinking of late (Banuri 1990; Braidotti *et al.* 1994; Goetz 1991; Harcourt 1994; Nandy 1989; Parpart 1993; Peet and Watts 1993; Porter *et al.* 1991; Schuurman 1993a; Watts 1993).

9   The Damongo project in Ghana, for example, ignored the physical characteristics of the Damongo area in order to "create" a feasible project. The land was a virtual wasteland, infested with tsetse flies and river blindness. As Grischow points out, "Colonial officials played down these problems by constructing the region as a place where Western technology could "rescue" the area . . . " (Grischow 1993: 3) (see also Armstrong 1987; Nindi 1990; Roe 1991).

10   This is still going on in the South. Nindi claims that "Foreign experts continue to exercise a virtual monopoly in the production of strategic regional, agricultural, industrial and urban plans only to name a few, while African planners fulfil the more routine tasks of annual planning, etc.." (1990: 42). Current structural adjustment programs continue the dominance of Western technical "fixes" as well (Barriteau Foster 1993a; Palmer 1991).

11   Although this process varied dramatically. The World Bank did not take the gender issue seriously until the late 1980s (World Bank 1993). For example, a 1989 report of the Bank declared WID "still a new field. . . . This paper is only a first step in identifying ways to improve women's opportunities" (World Bank 1989c).

12   Moser has outlined the shifts in mainstream development policy (Moser 1993; see introduction). Currently the ravages of economic decline and social service reductions of SAP initiatives has spawned a renewed interest

241

in human resource development (UNDP 1993). Democratization and civil society have also become women's issues (Hirschmann 1993; Nzomo 1992).

13 There are exceptions of course. CIDA has recently stated its commitment to fundamental change in gender relations in its WID policy (CIDA 1992; see chapter 11). However, the degree to which mainstream economic development models are challenged remains uncertain. A recent internal World Bank study of gender and economic adjustment policies, while critical of many Bank initiatives towards women, nevertheless concluded by emphasizing that that "there is at root no disagreement with what are identified as major factors essential for growth-oriented adjustment and poverty reduction . . . " (World Bank 1993: 26).

14 Small income-generating projects became the staple of nongovernmental organizations, but they were funded by some mainstream development agencies as well after the adoption of the basic human needs approach in the 1970s.

15 This is not to underestimate the importance of reliable data on women in the South; it is a warning about how much of this data is used.

16 Monica Opole, in her article "Revalidating Women's Knowledge on Indigneous Vegetables" reports that for the most part "Local capacities and knowledge have not been recognized or incorporated into mainstream development processes. . . . Development is still very top down . . ." (1993: 157).

17 In 1985, for every female CIDA-supported technical assistant abroad, there were 2.58 men. In 1989, the ratio had dropped to one woman to every 2.92 men. Scholarships are usually offered at the postgraduate or undergraduate level, which works against female applicants (Munday 1992).

18 The World Bank report on Kenya, for example, mentions women and education, agriculture, health, rural households and water supplies. It ignores violence against women, women's political roles and the need for fundamental changes in gender relations (World Bank 1989b).

19 Of course not all NGOs are in the alternative camp. DAWN classifies NGOs into seven types. Service-oriented NGOs and those affiliated with political parties are usually associated with mainstream, top-down development (examples are YWCA, Girl Guides, Women's Bureaus). Worker-based organizations and outside-initiated organizations are more directly concerned with the poor. Grassroots organizations arise out of the conditions and struggles of the poor and they adopt what is increasingly known as the empowerment approach to development. Research organizations can be elitist, but many are concerned with grassroots issues as well (such as Women and the Law in Southern Africa Research Project, WLSA). DAWN also mentions coalitions of women's organizations which have sprung up to fight a particular issue (Moser 1993: 198–203).

20 Of course, much of the language (and projects) aimed at minority groups in the North exhibit the same characteristics (see chapters 7–9).

21 Caroline Moser (1989; 1993) has defined the empowerment approach as largely Third World feminist writing and grassroots organization. According to her, it arose as a result of dissatisfaction with the equity approach to women's develoment. While initial writings in this vein contributed to and drew on socialist feminism, more recent writings have begun to draw on postcolonial/neo-colonial writings and postmodern feminist thought (see chapter 1).

22 Indeed, Aihwa Ong discusses an emerging counter-definition of modernity based on Chinese development models (Ong 1993).

GLOSSARY

Dichotomy refers to the ordering of concepts in binary, oppositional pairs, whereby the first term is superior to the second (e.g. masculine/feminine).

Discourse is a historically, socially and institutionally specific structure of statements, terms, categories and beliefs. It is considered the site where meanings are contested and power relations are determined.

Discursive practices produce and reproduce frameworks about the world so as to make it coherent, understandable and identifiable (see also *discourse*).

Dualism (see *dichotomy*).

Empowerment can be defined as a positive expression of power. It is distinguished from coercive interpretations of power in that it stresses the "enabling aspect" of power. Through cooperation and coordination people can use their capacity more productively and develop their potential more fully.

Enlightenment (thought) is a shorthand term for describing the major characteristics of Western knowledge and philosophy since the eighteenth century. It is usually associated with the belief in progress or *modernity*, dualistic or *dichotomous* thinking, making a sharp distinction between objective reality and subjective interpretation, and the search for a single grand theory that can be used to explain the world.

Gender refers to the social construction of masculinity and femininity, not the biological characteristics of the sexes.

Gender and Development (GAD) is an approach within the Women and Development literature. It focuses in particular on the social construction of gender roles and relations. Since GAD assumes that the gendered divisions of labor and power are constructed and not a natural given, it sets out to transform gender roles and relations. Thus far, the GAD approach is most widely represented within the academic community. (See also *Women in Development*, *Women and Development*, *Women, Environment and Development*.)

Modernity is a term closely associated with Enlightenment thinking. It reflects a state of being that is usually contrasted with traditional societies, which are supposedly burdened by superstition and lack of individuality. Instead, modernity is characterized by rationality, individualism, a positivist or scientific orientation, and democratic values.

Modernization is a (Eurocentric) theory of development embedded in the assumptions of *Enlightenment thought* and *modernity*. Using the modern/traditional dichotomy as its starting-point, modernization theory develops a set of rules and prescriptions to be followed if traditional societies want to become modern. These prescriptions often assume a linear progression from traditional to modern society and emphasize urbanization, industrialization and the adoption of Northern institutions and values.

(Neo-)colonial discourse refers to the representation of the "Third World" in Western academic writings, literature, policies, media coverage etc. The representation is based upon an assumed dichotomy between the West and the Third World whereby the latter is depicted as inferior, traditional, in need of (emergency) assistance (i.e. "Africa"), non-democratic and/or potentially dangerous. Among the most salient characteristics of (neo-)colonial discourse are the lack of differentiation among Third World countries, i.e. "one description fits all," and the recurrent portrayal of men (swarthy, undemocratic, macho, violent/revolutionary, sometimes lazy) and women (tradition-bound, voiceless, uneducated, dependent).

Postmodernism is a movement which originated within the arts and architecture. It is now a term which encompasses various approaches, including discourse analysis, genealogy, deconstructionism, textuality. What binds postmodernists is their rejection of *modernity*; they question, for instance, the Western knowledge (systems), the social construction of dominant interpretations, rationality, and are concerned with forms of resistance and silenced voices. (See also *dichotomy, Enlightenment thought, (neo-)colonial discourse, poststructuralism*.

Poststructuralism is a movement within the social sciences and literature (linguistics) critical of the assumptions embedded in *Enlightenment thought*. It parallels (and is now often conflated with) *postmodernism*, which originated in the arts and architecture. Among other things, poststructuralists reject a rationalist world view that rests upon *dichotomies*, assumes an objective reality, and sets out to develop a grand theory. (See also *dichotomy, Enlightenment thought, postmodernism*.)

Practical gender interests are usually articulated within the context of women's concrete living conditions, and tend to be related to their immediate problems. The term practical gender interests is often associated with the WID approach as well as with feminine movements and the private domain.

Praxis refers to the notion that there is no real distinction between theory and (political) action. They are interrelated and go hand in hand.

Strategic Gender Interests are usually articulated to reflect women's long-term needs and address existing gender hierarchies. The term is usually associated with the GAD approach as well as with feminist movements.

Women and Development (WAD) emerged in the late 1970s as a response to the early WID approach within the Women and Development literature. Influenced by radical critiques of development and patriarchy WAD advocates emphasized the importance of women's autonomy (from patriarchal and capitalist structures) by calling for women-only projects. The WAD approach is strongly represented within the NGO community. (See also *Gender and Development, Women in Development, Women, Environment and Development*.)

Women in Development (WID) is the earliest approach within the Women and Development literature. It originated as a feminist critique of dominant development modes, which often ignored women and deprived them of their traditional status and economic opportunities. Influenced by liberalism, the WID approach seeks to improve the situation of women by integrating (sometimes known as "mainstreaming") them into development policy and practice. (See also *Gender and Development, Women and Development, Women Environment and Development*.)

Women, Environment and Development (WED) is an approach within the Women and Development literature which emphasizes the intersections between gender, development and the environment. There are two strands to the WED approach. The first advocates a woman-centred approach to sustainable development based on the special relationship between women and nature. The second criticizes the modernist, growth-oriented (dominant) development model and its implications for women and the environment. (See also *Gender and Development, Women in Development, Women and Development.*)

BIBLIOGRAPHY

Abadan-Unat, N. (1984) "International Labour Migration and Its Effects upon Women's Occupation and Family Rules: A Turkish View," in UNESCO, *Women on the Move*, Paris: UNESCO.

Afshar, H. (ed.) (1991) *Women, Development and Survival in the Third World*, London: Longman.

Agarwal, B. (1991) *Engendering the Environment Debate: Lessons from the Indian Subcontinent*, CASID Distinguished Speaker Series, No. 8, East Lansing, MI: Center for Advanced Study of International Development, Michigan State University.

Ahmed, L. (1982) "Western Ethnocentrism and Perceptions of the Harem," *Feminist Studies*, 8: 521–34.

Aissou, A. (1987) *Les Beurs, l'école et la France*, Paris: Centre d'Information et d'Etudes sur les Migrations Internationales.

Alatas, S.F. (1993) "On the Indigenization of Academic Discourse," *Alternatives* 18: 307–38.

Alcoff, L. (1988) "Cultural Feminism versus Poststructuralism: The Identity Crisis in Feminist Theory," *SIGNS* 13, 3: 405–36. Reprinted in M.R. Malson, J.F. O'Barr, S. WestPhal-Wihl and M. Wyer (eds) (1989) *Feminist Theory in Practice and Process*, Chicago: University of Chicago Press.

Allen, P.G. (1986) *The Sacred Hoop: Recovering the Feminine in American Indian Traditions*, Boston: Beacon Press.

Alloula, M. (1986) *The Colonial Harem*, Minneapolis: University of Minnesota Press.

Althusser, L. (1971) "Ideology and Ideological State Apparatuses (Notes towards an Investigation)," in *Lenin and Philosophy and Other Essays*, translated by Ben Brewster, New York and London: Monthly Review Press.

Alvarez, S.E. (1990) *Engendering Democracy in Brazil*, Princeton, NJ: Princeton University Press.

Amin, S. (1974) *Accumulation on a World Scale: A Critique of the Theory of Under-development*, New York: Monthly Review Press.

—— (1977) *Imperialism and Unequal Development*, New York: Monthly Review Press.

Andersen, C. and Baud, I. (1987) "International Evolution of Women and Development Issue: UN and OECD-DAC," in C. Andersen and I. Baud (eds) *Women in Development Cooperation: Europe's Unfinished Business*, Centre for Development Studies, Antwerp, Technical University of Eindhoven, EADI Book Series 6, Tilburg, Netherlands: EADI.

247

BIBLIOGRAPHY

Anderson, P. (1986) "Conclusion WICP," *Social and Economic Studies* 25, 2: 291–324.

Andre, M. (1991) "L'Intégration au Féminin," *Hommes et Migrations* 1141 (March), 3.

Angelou, M. (1975) *Oh Pray My Wings Are Gonna Fit Me Well And Still I Rise*, New York: Random House.

Ankersmit, F.R. (1990) "Reply to Professor Zagorin," *History and Theory* 29, 3: 275–96.

Anthias, F. and Yuval-Davis, N. (1990) "Contextualising Feminism – Gender, Ethnic and Class Divisions," in T. Lovell (ed.) *British Feminist Thought: A Reader*, Oxford: Basil Blackwell.

Antrobus, P. (1989) "Gender Implications of the Development Crisis," in G. Beckford and N. Girvan (eds) *Development in Suspense: Selected Papers and Proceedings of the First Conference of Caribbean Economists*, Mona: Friedrich Ebert Stiftung.

Anwar, M. (1979) *The Myth of Return: Pakistanis in Britain*, London: Heinemann.

——— (1986) *Race and Politics: Ethnic Minorities and the British Political System*, London: Tavistock.

Anzaldua, G. (ed.) (1990) *Making Face, Making Soul/Haciendo Caras: Creative and Critical Perspective by Women of Color*, San Francisco: Aunt Lute Press.

Ardayfio-Schandorf, E. (1993) "Women and Forest Resources Management in the Northern Region of Ghana," Women, Environment and Development Network (WEDNET) Final Report, Nairobi: Environment Liaison Centre International (ELCI).

Arizpe, L. (1977) "Women in the Informal Labor Sector: The Case of Mexico City," *SIGNS* 3, 1: 25–37

Armstrong, A. (1987) "Tanzania's Expert-led Planning: An Assessment," *Public Administration and Development* 7: 261–71.

Armstrong, J. (1989) "Cultural Robbery, Imperialism: Voices of Native Women," *Trivia: A Journal of Ideas: Part II: Language/Differences: Writing in Tongues* 14: 21–3.

Asian Women Writers' Workshop (1988) *Right of Way*, London: Women's Press.

Assiter, A. (1990) *Althusser and Feminism*, Winchester, MA: Pluto Press.

Association of African Women for Research and Development (AAWORD) (1986) *Seminar on Research on African Women: What Type of Methodology*, Dakar: Council for the Development of Economic and Social Research in Africa (CODESRIA).

Bakhtin, M.M. (1981) *The Dialogic Imagination*, Austin, TX: University of Texas Press.

Bald, S.R. (1991) "Images of South Asian Migrants in Literature: Differing Perspectives," *New Community* 17, 3: 413–31.

Bandarage, A. (1984) "Women in Development : Liberalism, Marxism and Marxist Feminism," *Development and Change* 15: 495–515.

Bannerji, H. (1991) "But Who Speaks for Us? Experience and Agency in Conventional Feminist Paradigms," in H. Bannerji, L. Carty, K. Dehli, S. Heald and K. McKenna (eds) *Unsettling Relations: The University as a Site of Feminist Struggles*, Toronto: Women's Press.

Banuri, T. (1990) "Development and the Politics of Knowledge" and "Modernization and Its Discontents," in F. Marglin and S. Marglin (eds) *Dominating Knowledge: Development, Culture and Resistance*, Oxford: Clarendon Press.

Barbados, Government of (1988) *Barbados Development Plan 1983–88*, Bridgetown, Barbados: Government Printing Office.

────── (1989a) *Barbados Development Plan 1988–93: A Share For All*, Bridgetown, Barbados: Government Printing Office.

────── (1989b) "Structural Adjustment – Cure or Curse: Implications for Caribbean Development," paper presented to the Caribbean Development Bank Outreach Programme.

Barrett, M. (1980) *Women's Oppression Today: Problems in Marxist Feminist Analysis*, London: Verso.

────── (1985) "Introduction," in F. Engels *The Origin of the Family, Private Property and the State*, Middlesex: Penguin Books.

Barrios de Chungara, D. with M. Viezzer (1978) *Let Me Speak!* New York: Monthly Review Press.

Barriteau, E. (1992) "The Construct of a Postmodernist Feminist Theory for Caribbean Social Science Research," *Social and Economic Studies* 41, 2: 1–43.

────── (1993a) "A Feminist Perspective on Structural Adjustment Policies in the Caribbean to Development," paper presented to International Development Conference, Washington, DC, 11–13 January.

────── (1993b) "Gender and Development Planning in Barbados: 1961–1993," paper presented at Women and Development Seminar, Centre for Gender and Development Studies, University of the West Indies, Mona, 24 November.

────── and Clarke, R. (1989) "Grenadian Perceptions of the PRG and its Policies," *Social and Economic Studies* 38, 3: 53–92.

Barrow, C. (1983) *Guidelines for the Conduct of Social Surveys in the Caribbean: The Experience of a Five Island Interdisciplinary Survey*, Occasional Paper No. 17, Barbados: Institute for Social and Ecomomic Studies, EC, University of the West Indies.

────── (1986a) "Finding the Support: Strategies for Survival," *Social and Economic Studies* 35, 2: 131–76.

────── (1986b) "Male Images of Women in Barbados," *Social and Economic Studies* 35, 3: 51–64.

Batezat, E. and Mwalo, M. (1989) *Women in Zimbabwe*, Harare: Southern African Political Economy Series (SAPES) Trust.

Baudelot, C. and Establet, R. (1971) *L'Ecole capitaliste*, Paris: Maspéro.

Baudrillard, J. (1975) *The Mirror of Production*, St. Louis, MO: Telos Press.

Bauer, P.T. (1984) "Remembrance of Studies Past: Retracing First Steps," in G. M. Meier and D. Seers (eds) *Pioneers in Development*, Washington, DC: World Bank.

Beckles, H. McD. (1989a) *Natural Rebels: A Social History of Enslaved Women in Barbados*, London: Zed Books.

────── (1989b) *Corporate Power in Barbados. A Mutual Affair: Economic Injustice in a Political Democracy*, Bridgetown, Barbados: Lighthouse Publications.

Beckley, S. (1989) "Women as Agents/Recipients of Development Assistance: The Sierra Leone Case," in AAWORD (ed.) *Women as Agents and Beneficiaries of Development Assistance*, Occasional Paper Series No. 4, Dakar, Senegal: Association of African Women for Research and Development.

Begag, A. and Chaouite, A. (1990) *Ecarts d'Identité*, Paris: Editions du Seuil.

Behar, R. (1990) "Rage and Redemption: Reading the Life Story of a Mexican Marketing Woman," *Feminist Studies* 16, 2 (Summer): 223–58.

Beneria, L. and Sen, G. (1981) "Accumulation, Reproduction and Women's Role in Economic Development: Boserup Revisited," *SIGNS* 7, 2: 279–98.

Beneria, L. and Feldman, S. (eds) (1992) *Unequal Burden: Economic Crises, Persistent Poverty and Women's Work*, Boulder, CO: Westview Press.

Benjamin, M. (ed. and trans.) (1989) *Don't Be Afraid Gringo*, New York: Harper & Row.

Berlin, I. (1971) *Four Essays on Liberty*, Oxford: Oxford University Press.

Beverley, J. and Zimmerman, M. (1990) *Literature and Politics in the Central American Revolutions*, Austin, TX: University of Texas Press.

Bhabha, H.K. (1983) "The Other Question . . . Homi K. Bhabha Reconsiders the Stereotype and Colonial Discourse," *Screen* 24, 6: 18–36.

—— (1990) *Nation and Narration*, New York and London: Routledge.

—— (1994) *The Location of Culture*, New York and London: Routledge.

Blaney, D. and Inayatullah, N. (1994) "Prelude to a Conversation of Cultures in International Society? Todorov and Nandy on the Possibility of Dialogue," *Alternatives* 19: 23–51.

Boesveld, M., Helleman, C., Postel-Coster, E. and Schrijvers, J. (1986) *Towards Autonomy for Women: Research and Action to Support a Development Process*, Working Paper No. 1, The Hague: Raad van Advies voor het Wetenschappelijk Onderzoek in het kader van Ontwikkelingssamenwerking (RAWOO). [Formal advisory council to the Dutch government.]

Booth, D. (1985) "Marxism and Development Sociology: Interpreting the Impasse," *World Development* 13, 7: 761–87.

Bordo, S. (1990) "Feminism, Postmodernism, and Gender-Scepticism," in L. Nicholson (ed.) *Feminism/Postmodernism*, New York: Routledge, Chapman & Hall.

Boserup, E. (1970) *Woman's Role in Economic Development*, New York: St. Martin's and George Allen & Unwin.

Boulot, S. and Boyzon-Fradet, D. (1988) *Les Immigrés et l'école: Une course d'obstacles*, Paris: L'Harmattan et Centre d'Information et d'Etudes sur les Migrations Internationales.

Bourdieu, P. and Passeron, J.C. (1964) *Les Heritiers*, Paris: Editions de Minuit.

Bourne, J. (1983) "Towards an Anti-Racist Feminism," *Race and Class* 25: 1–22.

Bourque, S.C. (1989) "Gender and the State: Perspectives from Latin America," in S.E. Charlton, J. Everett and K. Staudt (eds) *Women, the State, and Development*, Albany, NY: State University of New York Press.

—— and Warren, K.B. (1990) "Access is Not Enough: Gender Perspectives on Technology and Education," in I. Tinker (ed.) *Persistent Inequalities: Women and World Development*, New York: Oxford University Press.

Bowles, G. and Duelli-Klein, R. (1983) "Introduction: Theories of Women's Studies and the Autonomy/Integration Debate" in G. Bowles and R. Duelli-Klein (eds) *Theories of Women Studies*, London: Routledge & Kegan Paul.

Braidotti, R., Charkiewicz, E., Häusler, S. and Wieringa, S. (eds) (1994) *Women, Environment and Sustainable Development: Towards a Theoretical Synthesis*, London: Zed Books in association with International Research and Training Institute for the Advancement of Women (INSTRAW).

Braxton, G. (1991) "Designing Projects as if Gender Mattered," paper presented at the Canadian Association of African Studies Conference, Toronto, 16–18 May.

Brereton, B. (1988) "General Problems and Issues on Studying the History of Women" in P. Mohammed and C. Shepherd (eds) *Gender in Caribbean Development*, Barbados: Women and Development Studies Project, University of the West Indies.

Brodber, E. (1982) *Perceptions of Caribbean Women*, Barbados: Institute for Social and Economic Research, EC, University of the West Indies.

—— (1986) "Afro-Jamaican Women at the Turn of the Century," *Social and Economic Studies* 35, 3: 23–50.

Brodribb, S. (1992) *Nothing Mat(t)ers: A Feminist Critique of Postmodernism*, Toronto: James Lorimer & Co.

Bromley, R. (1978) "Introduction – The Urban Informal Sector: Why Is It Worth Discussing?" *World Development* 6, 9/10: 1033–9.

Burgos-Debray, E. (ed.) (1984) *I . . . Rigoberta Menchu*, New York: Verso.

Butler, J. (1992) "Contingent Foundations: Feminism and the Question of 'Postmodernism'," in J. Butler and J.W. Scott (eds) *Feminists Theorize the Political*, New York and London: Routledge.

—— and Scott, J.W. (eds) (1992) *Feminists Theorize the Political*, New York and London: Routledge.

Buvinic, M. (1986) "Projects for Women in the Third World: Explaining their Misbehaviour," *World Development* 14, 5: 653–64.

—— (1989) "Investing in Poor Women: the Psychology of Donor Support," *World Development* 17, 7: 1045–57.

—— and Youssef, N.H. with von Elm, B. (1978) *Women Headed Households: The Ignored Factor in Development Planning*, Washington, DC: International Center for Research on Women.

Callaway, H. (1987) *Gender, Culture and Empire: European Women in Colonial Nigeria*, Urbana, IL: University of Illinois Press.

Callinicos, A. (1989) *Against Postmodernism*, Oxford: Polity Press.

Campbell, M. (1983) *Halfbreed*, Toronto: Goodread Biographies. Originally published by McClelland & Stewart, 1973. Published in a Seal Books edition in 1979.

Canadian International Development Agency (CIDA), Women in International Development Unit (1987) *WID Policy Framework*, Ottawa: CIDA.

——, Policy Branch (1991) *Sustainable Development*, Discussion Paper. Ottawa: CIDA.

——, WID Unit (1992) *CIDA, Women in Development Policy Framework*, Ottawa: CIDA.

Canning, K. (1994) "Feminist History after the Linguistic Turn: Historicizing Discourse and Experience," *SIGNS* 19, 2: 368–404.

Carloni, A.S. (1987) "Women in Development: AID's Experience 1973–1985," Vol. 1 Synthesis Paper, AID Program Evaluation Report No. 18, Washington, DC: Agency for International Development.

Castles, S. (1980) "The Social Time-bomb: Education of an Underclass in West Germany," *Race and Class* 21, 4: 369–87.

—— and Kosack, G. (1985) (first published, 1973) *Immigrant Workers and Class Structure in Western Europe*, New York: Oxford University Press.

Chatterjee, M. (1993) "Struggle and Development: Changing the Reality of Self-employed Workers," in G. Young, V. Samarasinghe and K. Kusterer (eds) *Women at the Center: Development Issues and Practices for the 1990s*, West Hartford, CT: Kumarian Press.

Chen, M. (1983) *A Quiet Revolution: Women in Transition in Rural Bangladesh*, Cambridge, MA: Shenkman.

Chhachhi, A. (1988) "Concepts in Feminist Theory: Consensus and Controversy," in P. Mohammed and C. Shepherd (eds) *Gender in Caribbean Development*, Barbados: Women and Development Studies Project, University of The West Indies.

Chimedza, R. (1993) "Women, Household Food Security and Wildlife

Resources," WEDNET Final Report, Nairobi: Environment Liaison Centre International (ELCI).

Chinchilla, N.S. (1977) "Industrialization, Monopoly Capitalism and Women's Work in Guatemala," *SIGNS* 3, 1: 38–56.

Chinemana, F. (1987) "Women and the Co-operative Movement in Zimbabwe," Report for the Ministry of Co-operative Development, Canadian University Students Overseas (CUSO), NOVIB, Harare: Government Printer.

Chow, R. (1991a) "CIDA's Women in Development Program: Evaluation Assessment Report," Ottawa: Canadian International Development Agency.

—— (1991b) "Violence in the Other Country: China as Crisis, Spectacle, and Woman," in C. Mohanty, A. Russo and L. Torres (eds) *Third World Women and the Politics of Feminism*, Bloomington, IN: Indiana University Press.

—— (1992) "Postmodern Automatons," in J. Butler and J.W. Scott (eds) *Feminists Theorize the Political*, New York and London: Routledge.

—— (1993) *Writing Diaspora*, Bloomington, IN: Indiana University Press.

Chowdhry, G. (1992) "Engendering Development? Theoretical Considerations for Women in Development," *International Journal of Humanities and Peace* Spring 1992

—— (1993) "Women and the International Political Economy," in F. D'Amico and P. Beckman (eds) *Women and World Politics*, South Hadley, MA: Bergin & Garvey.

—— (forthcoming) *International Financial Institutions, the State and Women Farmers in the Third World*, London: Macmillan

Cixous, H. (1981) "Castration or Decapitation?," translated by A. Kuhn, *SIGNS* 7, 1: 41–55.

Clarke, E. (1957) *My Mother Who Fathered Me*, London: George Allen & Unwin.

Clarke, R. (1986) "Women's Organization, Women's Interests," *Social and Economic Studies* 35, 3: 107–56.

Clifford, J. and Marcus G. (eds) (1986) *Writing Culture*, Berkeley: University of California Press.

Cocks, J. (1989) *The Oppositional Imagination: Critique and Political Theory*, New York: Routledge.

Collins, J.L. (1991) "Women and the Environment: Social Reproduction and Sustainable Development," in R.S. Gallin and A. Ferguson (eds) *The Women and International Development Annual, Vol. 2*, Boulder, CO: Westview Press.

Collins, P.H. (1989) "The Social Construction of Black Feminist Thought," *SIGNS* 14, 4: 745–73.

—— (1990) *Black Feminist Thought: Knowledge, Consciousness, and the Politics of Empowerment*, Boston: Unwin Hyman.

Commission of the European Community (1990) "Thematic Evaluation on the Integration of Women in Rural Development," Brussels: CEC.

Commonwealth Expert Group on Women and Structural Adjustment (1989) *Engendering Adjustment for the 1990s* (report), London: Commonwealth Secretariat.

Comte, A. (1975) *August Comte and Positivism: Essential Writings*, New York: Harper & Row.

Connelly, P., Li, T., MacDonald, M. and Parpart, J. (forthcoming) "Restructured Worlds/Restructured Debates: Globalization, Development and Gender," *Canadian Journal of Development Studies*.

Cooper, F. (1981) "Africa and the World Economy," *African Studies Review* 24, 2/3: 1–86.

Corbridge, S. (1989) "Post-Marxism and Development Studies: Beyond the Impasse," *World Development* 8, 5: 623–40.

Cornell, D. (1992) "Feminist Legal Reform" (mimeo).

Cornia, G.A., Jolly, R. and Stewart, F. (eds) (1987) *Adjustment with a Human Face*, Vol. I, *Protecting the Vulnerable and Promoting Growth*, Oxford: Oxford University Press.

Coulibally, S. (1993) "Women, Migration, and the Management of Natural Resources," WEDNET Final Report, Nairobi: Environment Liaison Centre International (ELCI).

Courteau, J. (1974) "The Image of Women in the Novels of Graciliano Ramos," *Review Interamericana* 4, 2: 162–76.

Crapanzano, V. (1984) "Life Histories," *American Anthropologist* 86, 4 (December): 953–9.

Crush, J. (ed.) (1994) *Development Discourse*, London: Routledge.

Cuales, S. (1988) "Some Theoretical Considerations on Social Class, Class Consciousness and gender consciousness," in P. Mohammed and C. Shepherd (eds) *Gender in Caribbean Development*, Barbados: Women and Development Studies Project, University of The West Indies.

Culler, J. (1982) *On Deconstruction: Theory and Criticism after Structuralism*, Ithaca, NY: Cornell University Press.

Currie, D. and Kazi, H. (1987) "Academic Feminism and the Process of De-Radicalization: Reexamining the Issues," *Feminist Review* 25, 77–98.

Curtin, P. (1974) *The Image of Africa*, Madison, WI: Wisconsin University Press.

Daly, M. (1978) *Gyn/Ecology: The Metaethics of Radical Feminism*, Boston: Beacon Press.

Dankelman, I. and Davidson, J. (1988) *Women and Environment in the Third World: Alliance for the Future*, London: Earthscan.

Dauber, R. and Cain, M. (eds) (1981) *Women and Technological Change in Developing Countries*, Boulder, CO: Westview Press.

DAWN (1991) *Alternatives*, vols. I and II, Rio de Janeiro: Development Alternatives with Women for a New Era.

de Beauvoir, S. (1949) "The Point of View of Historical Materialism," in S. de Beauvoir *The Second Sex*, London: Penguin Books.

de Boef, W., Amanor, K. and Wellard, K. with Bebbington, A. (eds) (1993) *Cultivating Knowledge: Genetic Diversity, Farmer Experimentation and Crop Research*, London: Intermediate Technology Publications.

de Groot, J. (1991) "Conceptions and Misconceptions: The Historical and Cultural Context of Discussion on Women and Development," in H. Afshar (ed.) *Women, Development and Survival in the Third World*, London: Longman.

de Lauretis, T. (1984) *Alice Doesn't*, Bloomington, IN: Indiana University Press.

—— (1990) "Eccentric Subjects: Feminist Theory and Historical Consciousness," *Feminist Studies* 16, 1: 115–50.

Deleuze, G. (1990) *Pourparlers*, Paris: Minuit.

Delphy, C. (1984) "A Materialist Feminism is Possible" in C. Delphy *Close to Home: A Materialist Analysis of Women's Oppression*, Amherst, MA: University of Massachusetts Press.

Department of Indian and Northern Affairs, Canada (1981) *Indian Acts and Amendments, 1868–1950*, 2nd edition, Ottawa: Government Printing Office.

Derrida, J. (1976) *Of Grammatology*, translated by Gayatri Spivak, Baltimore, MD: Johns Hopkins University Press.

di Leonardo, M. (ed.) (1991) *Gender at the Crossroads of Knowledge*, Berkeley: University of California Press.

Di Stefano, C. (1990a) "Dilemmas of Difference: Feminism, Modernity, and Postmodernism," in L. Nicholson (ed.) *Feminism/Postmodernism*, New York: Routledge, Chapman & Hall.

——— (1990b) "Masculine Marx," in M. Lyndon Shanley and C. Pateman (eds) *Feminist Interpretations and Political Theory*, Oxford: Polity Press.

Diamond, I. and Quinby, L. (1988) *Feminism and Foucault*, Boston: Northeastern University Press.

Dixon, R. (1978) *Rural Women at Work: Strategies for Development in South Asia*, Baltimore, MD: Johns Hopkins University Press.

Djebar, A. (1985) *L'Amour, la fantasia*, Paris: Editions J.C. Lattes.

Drinkwater, M. (1991) *The State and Agrarian Change in Zimbabwe's Communal Areas*, New York: St. Martin's Press.

DuBois, M. (1991) "The Governance of the Third World: A Foucauldian Perspective on Power Relations in Development," *Alternatives* 16, 1: 1–30.

Duncan, N.O.K. (1989) *Women and Politics*, Barbados: Institute for Social and Economic Research, EC, University of the West Indies.

Durant-Gonzales, V. (1982) "The Realm of Female Familial Responsibility" in *Women and the Family*, Barbados: Institute for Social and Economic Research, EC, University of the West Indies.

Dwyer, K. (1982) *Moroccan Dialogues*, Baltimore, MD: Johns Hopkins University Press.

Eagleton, T. (1983) *Literary Theory*, Minneapolis: Minnesota University Press.

Edwards, M. (1989) "The Irrelevance of Development Studies," *Third World Quarterly* 11, 1: 116–35.

Eisenstein, H. and Jardine, A. (eds) (1988) *The Future of Difference*, New Brunswick, NJ: Rutgers University Press.

Eisenstein, Z. (1981) *The Radical Future of Liberal Feminism*, New York: Longman.

Ellis, P. (ed.) (1986) *Women of the Caribbean*, London: Zed Books.

Elshtain, J.B. (1981) *Public Man, Private Woman: Women in Social Thought*, Princeton: Princeton University Press.

Elson, D. (ed.) (1991) *The Male Bias in the Development Process*, Manchester: Manchester University Press.

Emberley, J.V. (1990–1) "'A Gift for Languages': Native Women and the Textual Economy of the Colonial Archive," *Cultural Critique* 17: 21–50. Reprinted in J.V. Emberley (1993) *Thresholds of Difference: Feminist Critique, Native Women's Writings, Postcolonial Theory*, Toronto: University of Toronto Press.

——— (1993) *Thresholds of Difference: Feminist Critique, Native Women's Writings, Postcolonial Theory*, Toronto: University of Toronto Press.

Enloe, C. (1989) *Bananas, Beaches and Bases: Making Feminist Sense of International Politics*, Berkeley: University of California Press.

Escobar, A. (1984–5) "Discourse and Power in Development: Michel Foucault and the Relevance of His Work to the Third World," *Alternatives* 10, 3: 377–400.

——— (1992) "Imagining a Post-Development Era? Critical Thought, Development and Social Movements," *Social Text* 31/32: 20–56.

Esteva, G. (1987) "Regenerating People's Space," *Alternatives* 12, 1: 125–52.

Evans, M. (1982) "In Praise of the Theory: The Case For Women's Studies," in G. Bowles and R. Duelli-Klein (eds) *Theories of Women Studies*, London: Routledge & Kegan Paul.

Fals Borda, O. and Rahman, A. (eds) (1991) *Action and Knowledge: Breaking the Monopoly with Participatory Action-Research*, New York: Apex Press.

Fanon, F. (1963) *The Wretched of the Earth*, New York: Grove Press.

Felski, R. (1989) "Feminism, Postmodernism, and the Critique of Modernity," *Cultural Critique*, Fall: 33–56.

Ferguson, J. (1985) "The Bovine Mystique: Power, Property and Livestock in Rural Lesotho," *MAN* 20, 4: 647–74.

———— (1990) *The Anti-Politics Machine*, Cambridge: Cambridge University Press.

Ferguson, K.E. (1988) "Subject Centeredness in Feminist Discourses," in K.B. Jones and A.G. Jonasdottir (eds) *The Political Interests of Gender*, London: Sage Publications.

———— (1991) "Interpretation and Genealogy in Feminism," *SIGNS* 16, 2: 322–39.

———— (1993) *The Man Question: Visions of Subjectivity in Feminist Theory*, Berkeley: University of California Press.

Flax, J. (1987) "Postmodernism and Gender Relations in Feminist Theory," *SIGNS* 12, 4: 621–43. Reprinted (1989) in M.R. Malson, J.F. O'Barr, S. WestPhal-Wihl and M. Wyer (eds) *Feminist Theory in Practice and Process*, Chicago: University of Chicago Press.

———— (1990a) "What is Enlightenment: A Feminist Rereading," paper prepared for Conference on Postmodernism and the Rereading of Modernity, University of Essex, 9–11 July.

———— (1990b) *Thinking Fragments: Psychoanalysis, Feminism, and Postmodernism in the Contemporary West*, Berkeley: University of California Press.

———— (1992) "Feminists Theorize the Political," in J. Butler and J.W. Scott (eds) *Feminists Theorize the Political*, New York and London: Routledge.

Flora, C.B. (1983) "Incorporating Women into International Development Programs: The Political Phenomenology of a Private Foundation," in K.A. Staudt and J.S. Jaquette (eds) *Women in Developing Countries: A Policy Focus*, New York: Haworth.

Folbre, N. (1991) "The Unproductive Housewife: Her Evolution in Nineteenth-Century Economic Thought," *SIGNS* 16, 3: 463–84.

Fontana, A. and Pasquino, P. (1984) "Truth and Power," interview with Michel Foucault in P. Rabinow (ed.) *The Foucault Reader*, New York: Penguin Books.

Foucault, M. (1972) *The Archaeology of Knowledge and the Discourse on Language*, New York: Tavistock Publications & Harper Colophon.

———— (1973) *The Order of Things: An Archaeology of the Human Sciences*, New York: Vintage Books.

———— (1979) (published in French, 1975) *Discipline and Punish*, translated by S. Sheridan, New York: Vintage Books.

———— (1980) *Power/Knowledge: Selected Interviews and Other Writings, 1972–1977*, translated by C. Gordon, New York: Harvester Press.

———— (1986) "What Is Enlightenment?," in P. Rabinow (ed.) *The Foucault Reader*, Harmondsworth: Penguin Books.

Foweraker, J. (1990) "Popular Movements and Political Change in Mexico," in J. Foweraker and A.L. Craig (eds) *Popular Movements and Political Change in Mexico*, Boulder, CO: Lynne Rienner.

Frank, A.G. (1967) *Capitalism and Underdevelopment in Latin America: Historical Studies of Chile and Brazil*, New York: Monthly Review Press.

———— (1978) *Dependent Accumulation and Under-development*, London: Macmillan.

Fraser, N. (1989) *Unruly Practices: Power, Discourse and Gender in Contemporary Social Theory*, Oxford: Polity Press.

—— and Nicholson, L. (1990) "Social Criticism without Philosophy: An Encounter between Feminism and Postmodernism," in L. Nicholson (ed.) *Feminism/Postmodernism*, New York: Routledge, Chapman & Hall.

Friedman, M. (1962) *Capitalism and Freedom*, Chicago: University of Chicago Press.

Fukuyama, F. (1989) "The End of History?" *The National Interest* 16 (Summer): 3–18.

Fuss, D. (1989) *Essentially Speaking: Feminism, Nature and Difference*, New York: Routledge.

Gahlot, D. (1991) "'SPARC' of Hope for India's Slum Dwellers," *IDRC Reports* 19, 1: 8–9.

Garlake, P. (1982) "Prehistory and Ideology in Zimbabwe," *Africa* 52, 3: 1–20.

Gewertz, D. and Errington, F. (1991) "We Think, Therefore They Are? On Occidentalizing the World," *Anthropological Quarterly* 64, 2: 80–91.

Gillespie, K. (1989) "The Key to Unlocking Sustainable Development," Washington, DC: World Bank, mimeo.

Gilligan, C. (1982) *In a Different Voice*, Cambridge: Harvard University Press.

Goetz, A.M. (1988) "Feminism and the Limits of the Claim to Know: Contradictions in the Feminist Approach to Women in Development," *Millennium* 17,3: 477–96. Reprinted in R. Grant and K. Newland (eds) (1991) *Gender and International Relations*, Bloomington, IN: Indiana University Press.

Gosses, A. (n.d.) "Women in Micro- and Small-scale Enterprises: Some Views on Policy Integration" (mimeo).

Griffiths, L. and Campbell, M. (1989) *The Book of "Jessica": A Theatrical Transformation*, Toronto: Coach House Press.

Grischow, J. (1993) "Creating Underdevelopment in the Northern Territories of the Gold Coast: The Damongo Groundnut Scheme, 1947–1957," paper presented at the Canadian Association of African Studies Conference, Toronto, May.

Guha, R. and Spivak, G.C. (1988) *Selected Subaltern Studies*, Oxford: Oxford University Press.

Gutting, G. (1989) *Michel Foucault's Archaeology of Scientific Reason*, Cambridge: Cambridge University Press.

Hagen, E.E. (1962) *On the Theory of Social Change*, Illinois: Dorsey Press.

Harasym, S. (ed.) (1990) *The Post-Colonial Critic*, London: Routledge.

Haraway, D. (1990) "A Manifesto for Cyborgs: Science, Technology, and Socialist Feminism in the 1980s," in L. Nicholson (ed.) *Feminism/Postmodernism*, New York: Routledge, Chapman & Hall.

—— (1991) *Simians, Cyborgs, and Women: The Reinvention of Nature*, New York: Routledge.

Harcourt, W. (1994) *Feminist Perspectives on Sustainable Development*, London and New Jersey: Zed Books, in association with the Society for International Development (Rome).

Harding, S. (1986) *The Science Question in Feminism*, Ithaca, NY: Cornell University Press.

—— (ed.) (1987) *Feminism and Methodology*, Bloomington, IN: Indiana University Press.

—— (1989) "The Instability of the Analytical Categories of Feminist Theory," in M.R. Malson, J.F. O'Barr, S. WestPhal-Wihl and M. Wyer (eds) *Feminist Theory in Practice and Process*, Chicago: University of Chicago Press.

256

——— (1991) *Whose Science? Whose Knowledge?* Ithaca, NY: Cornell University Press.

——— (1992) "Subjectivity, Experience and Knowledge," *Development and Change* 23, 3: 175–94.

Hart, K. (ed.) (1989) *Women and the Sexual Division of Labour in the Caribbean*, University of West Indies, Mona Campus, Jamaica: Consortium Graduate School of Social Science Research.

Hartmann, H. (1981) "The Unhappy Marriage of Marxism and Feminism: Towards a More Progressive Union," in L. Sargent (ed.) *Women and Revolution: A Discussion of the Unhappy Marriage of Marxism and Feminism*, Boston: South End Press.

Hartsock, N. (1984) *Money, Sex and Power: Towards a Feminist Historical Materialism*, Boston: Northeastern University Press.

——— (1987) "The Feminist Standpoint: Developing the Ground for a Specifically Feminist Historical Materialism," in S. Harding (ed.) *Feminism and Methodology*, Bloomington, IN: Indiana University Press.

——— (1989) "Postmodernism and Political Change: Issues for Feminist Theory," *Cultural Critique* Winter: 15–33.

——— (1990) "Foucault on Power: A Theory for Women," in L. Nicholson (ed.) *Feminism/Postmodernism*, New York: Routledge, Chapman & Hall.

Harvey, D. (1989) *The Condition of Postmodernity: An Enquiry into the Origins of Cultural Change*, Oxford: Basil Blackwell.

Hawkesworth, M.E. (1989) "Knowers, Knowing, Known: Feminist Theory and Claims of Truth," *SIGNS* 14, 3: 533–57. Also in M.R. Malson, J.F. O'Barr, S. WestPhal-Wihl and M. Wyer (eds) (1989) *Feminist Theory in Practice and Process*, Chicago: University of Chicago Press.

Hekman, S.J. (1987) "The Feminization of Epistemology, Gender and Social Sciences," *Women and Politics* 7, 3: 7–25.

——— (1990) *Gender and Knowledge*, Boston: Northeastern University Press.

Hennessy, R. (1993) *Materialist Feminism and the Politics of Discourse*, New York: Routledge.

Herz, B. and Measham, A.R. (1987) *The Safe Motherhood Initiative: Proposals for Action*, Washington, DC: World Bank.

Heyzer, N., Riker, J. and Quizon, A.B. (eds) (1995) *Government–NGO Relations in Asia: Prospects and Challenges for People-Centred Development*, London: Macmillan.

——— and Wee, V. (forthcoming) "Gender, Environment and Development" (mimeo).

Hirsch, M. and Fox Keller, E. (eds) (1990) *Conflicts in Feminism*, New York: Routledge.

Hirschmann, D. (1993) "Democracy and Gender: A Practical Guide to USAID Programs," Washington, DC: US Agency for International Development.

Hirschmann, N.J. (1992) *Rethinking Obligation: A Feminist Method for Political Theory*, Ithaca, NY: Cornell University Press.

Hobart, M. (1993) "As I Lay Laughing Encountering Global Knowledge in Bali," Association of Social Anthropology (ASA), IV Decennial Conference, Oxford, England, 26–31 July.

Hoberman, L.S. (1974) "Hispanic American Women as Portrayed in the Historical Literature: Types or Archetypes?," *Review Interamericana* 4, 2: 136–47.

hooks, b. (1984) *Feminist Theory: From Margin to Center*, Boston: South End Press.

——— (1988) *Talking Back: Thinking Feminist, Thinking Black*, Boston: South End Press.

257

———— (1990) *Yearning: Race, Gender and Cultural Politics*, Boston: South End Press.

———— (1992) *Black Looks: Race, Representation*, Boston: South End Press.

Hosken, F. (1981) "Female Genital Mutilation and Human Rights," *Feminist Issues*, 1 (3).

Hull, G. T., Scott, P. and Smith, B. (eds) (1982) *All the Women are White, All the Blacks are Men, but Some of Us Are Brave: Black Women's Studies*, New York: Feminist Press.

Huntington, S. (1971) "Change to Change: Modernization, Development and Politics," *Comparative Politics*, 3,3: 283–322.

———— (1993) "The Clash of Civilizations?" *Foreign Affairs* 72, 3: 22–49.

Hutcheon, L. (1989) *The Politics of Postmodernism*, London: Routledge.

ISIS International (1986) *Women, Struggles and Strategies: Third World Perspectives*, Rome: ISIS International.

Jaggar, A. and Bordo, S. (eds) (1989) *Gender/Body/Knowledge*, New Brunswick, NJ: Rutgers University Press.

James, S.M. and Busia, A.P.A. (1993) *Theorizing Black Feminisms*, London: Routledge.

Jameson, F. (1991) *Postmodernism or, the Cultural Logic of Late Capitalism*, Durham, NC: Duke University Press.

Jamieson, K. (1986) "Sex Discrimination and the Indian Act," in J.R. Ponting (ed.) *Arduous Journey: Canadian Indians and Decolonization*, Toronto: McClelland & Stewart.

Jaquette, J. (1982) "Women and Modernization Theory: A Decade of Feminist Criticism," *World Politics* 34, 2: 267–84.

———— (ed.) (1989) *The Women's Movement in Latin America*, Boston: Unwin Hyman.

Jara, R. and Vidal, H. (eds) (1986) *Testimonio y Literatura*, Minneapolis, MN: Institute for the Studies of Ideologies and Literature.

Jayawardena, K. (1986) *Feminism and Nationalism in the Third World*, London: Zed Books.

Jaycox, E. (1993) "Capacity Building: The Missing Link in African Development," address to the African-American Institute Conference *African Capacity Building: Effective and Enduring Partnerships*, Reston, Virginia, 20 May (mimeo).

Jeffery, P. (1976) *Migrants and Refugees: Muslims and Christian Pakistani Families in Bristol*, Cambridge: Cambridge University Press.

Joekes, S. (1987) *Women in the World Economy*, Oxford: Oxford University Press.

Johnson-Odim, C. (1991) "Common Themes, Different Contexts: Third World Women and Feminism," in C. Mohanty, A. Russo and L. Torres (eds) *Third World Women and the Politics of Feminism*, Bloomington, IN: Indiana University Press.

Johnston, D. (1991) "Constructing the Periphery in Modern Global Politics," in C. Murphy and R. Tooze (eds) *The New International Political Economy*, Boulder, CO: Lynne Rienner.

Jones, K.B. and Jonasdottir, A.G. (eds) (1988) *The Political Interests of Gender*, London: Sage Publications.

Joseph, R. (1981) "The Significance of the Grenada Revolution to Women in Grenada," *Bulletin of Eastern Caribbean Affairs* 7, 1: 16–19.

Kabbani, R. (1986) *Europe's Myths of the Orient*, Bloomington, IN: Indiana University Press.

Kabeer, N. (1991) "Rethinking Development from a Gender Perspective: Some Insights from the Decade," paper presented at the Conference on Women and Gender in Southern Africa, University of Natal, Durban.

BIBLIOGRAPHY

Kandiyoti, D. (1990) "Women and Rural Development Policies: The Changing Agenda," *Development and Change* 21, 1: 5–22.

Kant, E. (1963) *On History*, L.W. Beck (ed.), New York: Bobbs Merrill.

Kardam, N. (1989) "Women and Development Agencies," in R. Gallin, M. Aronoff and A. Ferguson (eds) *The Women and International Development Annual*, vol. 1, Boulder, CO: Westview Press.

—— (1991) *Bringing Women In: Women's Issues in International Development Programs*, Boulder, CO: Lynne Rienner.

Kay, C. (1989) *Latin American Theories of Development and Underdevelopment*, Boulder, CO: Lynne Rienner.

Keesing, R.M. (1994) "Colonial and Counter-Colonial Discourse in Melanesia," *Critique of Anthropology* 14, 1: 41–58.

Kelly, D. (1987) *Hard Work, Hard Choices: A Survey of Women in St. Lucia's Export Oriented Electronics Factories*, Occasional Paper No. 20, Barbados: Institute for Social and Economic Studies, EC, University of the West Indies.

Kettel, B. (1993) "Gender and Environments: Lessons From WEDNET," unpublished paper, Toronto: Faculty of Environmental Studies, York University.

King, D. (1989) "Multiple Jeopardy, Multiple Consciousness: The Context of a Black Feminist Ideology," *SIGNS* 14, 1: 42–72. Reprinted in M.R. Malson, J.F. O'Barr, S. WestPhal-Wihl and M. Wyer (eds) (1989) *Feminist Theory in Practice and Process*, Chicago: University of Chicago Press.

Kiraitu, M. (1992) "The Freedom of Expression, Association and Assembly and the Politics of Non-violent Action," paper presented at the International Commission of Jurists' Kenya Section Seminar on Freedom of Expression, Association and Assembly, 6–9 May.

Kondo, D. (1990) *Crafting Selves*, Chicago: University of Chicago Press.

Kruger, B. (1984) *We Won't Play Nature to Your Culture*, London: Institute of Contemporary Arts.

Kruks, S. (1992) "Gender and Subjectivity: Simone de Beauvoir and Contemporary Feminism," *SIGNS* 18, 1: 89–110.

Landes, J.B. (1988) *Women in the Public Sphere: In the Age of The French Revolution*, Ithaca, NY: Cornell University Press.

Langness, L.L. and Frank, G. (1981) *Lives*, Novato, CA: Chandler & Sharp.

Lazreg, M. (1988) "Feminism and Difference: The Perils of Writing as a Woman on Women in Algeria," *Feminist Studies* 14, 1: 81–107.

Leach, M. (1991) "Traps and Opportunities: Some Thoughts on Approaches to Gender, Environment and Social Forestry, with Emphasis on West Africa," paper presented at the DSA Women in Development Study Group Meeting, Institute of Development Studies, University of Sussex, Sussex.

Leacock, E. (1980) "Montagnais Women and the Jesuit Program for Colonization," in M. Etienne and E. Leacock (eds) *Women and Colonization: Anthropological Perspectives*, New York: Praeger.

Leacock, E. and Safa, H. and contributors (1986) *Women's Work: Development and the Division of Labor by Gender*, South Hadley, MA: Bergin & Garvey.

Lerner, D. (1964) *The Passing of Traditional Society*, New York: Free Press.

Lévi-Strauss, C. (1963) "Language and the Analysis of Social Laws," in *Structural Anthropology*, translated by Claire Jacobson and Brooke Grundfest Schoepf, New York: Basic Books.

Lewis, W.A. (1966) *Development Planning: The Essentials of Economic Policy*, London: George Allen & Unwin.

—— (1978) *The Theory of Economic Growth*, London: George Allen & Unwin.

—— (1984) "Development Economics in the 1950s," in G.M. Meier and D. Seers (eds) *Pioneers in Development*, Washington, DC: World Bank.

Lindblom, C. (1965) *The Intelligence of Democracy*, New York: Free Press.

—— (1977) *Politics and Markets*, New York: Basic Books.

Lloyd, M. (1991) "Feminist 'Hyphenisation': Decentring the 'Prototypical Woman'," paper presented to the Women in a Changing Europe Conference, University of Alborg, Denmark.

Longfellow, B. (1992) "The Melodramatic Imagination in Quebec and Canadian Women's Feature Films," *Cineaction* 28: 48–56.

Lorde, A. (1981) "The Master's Tools Will Never Dismantle the Master's House," in C. Moraga and G. Anzaldua (eds) *This Bridge Called My Back: Writings by Radical Women of Color*, Watertown, MA: Persephone Press.

—— (1984) *Sister Outsider*, New York: Crossing Press.

Lovibond, S. (1990) "Feminism and Postmodernism," in R. Boyne and A. Rattansi (eds) *Postmodernism and Society*, New York: St. Martin's Press.

Lyotard, J.-F. (1984) *The Postmodern Condition: A Report on Knowledge*, Minneapolis: Minnesota University Press.

McAfee, K. (1991) *Storm Signals: Structural Adjustment and Development Alternatives in the Caribbean*, London: Zed Books.

McClintock, A. (1992) "The Angel of Progress: Pitfalls of the Term 'Post-Colonialism'," *Social Text* 31/32: 84–98.

McGuire, J. and Popkin, B. (1990) *Helping Women Improve Nutrition in the Developing World: Beating the Zero Sum Game*, World Bank Technical Paper No. 114, Washington, DC: World Bank.

Mackenzie, F. (1993) "Exploring the Connections: Structural Adjustment, Gender and the Environment," *Geoforum* 24, 1: 71–87.

Maclure, R. (forthcoming) "Primary Health Care and Donor Dependency: A Case Study of Nongovernmental Assistance in Burkina Faso," *International Journal of Health Services*.

McNay, L. (1992) *Foucault and Feminism*, London: Polity Press.

Maguire, P. (1984) *Women in Development: An Alternative Analysis*, Amherst, MA: Center for International Education, University of Massachusetts.

Mahoney, M.A. and Yngvesson, B. (1992) "The Construction of Subjectivity and the Paradox of Resistance: Reintegrating Feminist Anthropology and Psychology," *SIGNS* 18, 1: 44–73.

Malachlan, C.M. (1974) "Modernization of Female Status in Mexico: The Image of Women's Magazines," *Review Interamericana* 4, 2: 246–57.

Manzo, K. (1991) "Modernist Discourse and the Crisis of Development Theory," *Studies in Comparative International Development* 26, 2: 3–36.

Maracle, L. (1988) *I Am Woman*, North Vancouver, BC: Write-On Press.

—— (1989) "Moving Over," *Trivia: A Journal of Ideas: Part II: Language/Differences: Writing in Tongues* 14: 9–12.

Marchand, M.H. (1994) "Latin American Voices of Resistance: Women's Movements and Development Debates," in S.J. Rosow, N. Inayatullay and M. Rupert (eds) *The Global Political Economy as Political Space*, Boulder, CO: Lynne Rienner.

—— (1994) "Gender and New Regionalism in Latin America: Inclusion/Exclusion," *Third World Quarterly* (Special Issue: The South in the New World (Dis)Order) 15, 1: 63–76.

—— (1994) "The Political Economy of North–South Relations," in R. Stubbs and G.R.D. Underhill (eds) *Political Economy and the Changing Global Order*, London: Macmillan Press.

Marcus, G.E. and Fischer, M.M.J. (1986) *Anthropology as Cultural Critique*, Chicago: University of Chicago Press.

Mascarenhas, J. (1993) "Participatory Approaches to Management of Local Resources in Southern India," in W. de Boef, K. Amanor and K. Wellard with A. Bebbington (eds), *Cultivating Knowledge: Genetic Diversity, Farmer Experimentation and Crop Research*, London: Intermediate Technology Publications.

Mascia-Lees, F., Sharpe, P. and Cohen, C.B. (1989) "The Postmodernist Turn in Anthropology: Cautions from a Feminist Perspective," *SIGNS* 15, 1: 394–408.

Massiah, J. (1982a) "Indicators of Women in Development: A Preliminary Framework for the Caribbean," in *Women and the Family*, Barbados: Institute for Social and Economic Research, EC, University of the West Indies.

—— (1982b) "Women Who Head Households," in *Women and the Family*, Barbados: Institute for Social and Economic Research, EC, University of the West Indies.

—— (1986a) "Postscript. The Utility of WICP Research in Social Policy Formation," *Social and Economic Studies* 35, 3: 157–81.

—— (1986b) "Women in the Caribbean Project: An Overview," *Social and Economic Studies* 35, 2: 1–30.

—— (1986c) "Work in the Lives of Caribbean Women," *Social and Economic Studies* 35, 2: 177–240.

—— (1988) "Researching Women's Work 1985 and Beyond," in P. Mohammed and C. Shepherd (eds) *Gender in Caribbean Development*, Barbados: Women and Development Studies Project, University of the West Indies.

Massolo, A. (1992) *Por Amor y Coraje*, Mexico City, El: Colegio de México.

Mathema, C. with I. Staunton (1985) *Cooperatives: What about Them?* Harare: Ministry of Education, Government of Zimbabwe.

Mathur, G.V. (1989) "The Current Impasse in Development Thinking: the Metaphysic of Power," *Alternatives* 14: 463–79.

Mbilinyi, M. (1989) "'I'd Have Been a Man': Politics and the Labour Process in Producing Personal Narratives," in Personal Narratives Group, *Interpreting Women's Lives*, Bloomington, IN: Indiana University Press.

Meier, G.M. and Seers, D. (1984) *Pioneers in Development*, Washington, DC: World Bank.

Melucci, A. (1980) "The New Social Movements: A Theoretical Approach," *Social Science Information* 19, 2: 199–226.

Memmi, A. (1967) *The Colonizer and the Colonized*, Boston: Beacon Press.

Mernissi, F. (1975) *Beyond the Veil: Male and Female Dynamics in Modern Muslim Society*, Cambridge, MA: Schenkman. Revised edition (1987), Bloomington, IN: Indiana University Press.

Mies, M. (1982) *The Lace Makers of Narsapur: Indian Housewives Produce for the World Market*, London: Zed Books.

—— (1989) *Patriarchy and Accumulation on a World Scale: Women in the International Division of Labour*, London: Zed Books.

—— and Shiva, V. (1993) *Ecofeminism*, Halifax and London: Fernwood Publications and Zed Books.

Miller, E. (1991) *Men At Risk*, Kingston, Jamaica: Jamaica Publishing House.

Minh-ha, T.T. (1987) "Difference: 'A Special Third World Women Issue'," *Feminist Review* 25: 5–22.

—— (1989) *Women, Native, Other*, Bloomington, IN: Indiana University Press.

Moghadam, V.M. (1992) "Development and Women's Emancipation: Is There a Connection?" *Development and Change* 23, 3: 215–55.

Mohammed, P. and Shepherd, C. (eds) (1988) *Gender in Caribbean Development*, Barbados: Women and Development Studies Project, University of the West Indies.

Mohanty, C. (1988) "Under Western Eyes: Feminist Scholarship and Colonial Discourses," *Feminist Review* 30: 61–88; also (1991a) in C. Mohanty, A. Russo and L. Torres (eds) *Third World Women and the Politics of Feminism*, Bloomington, IN: Indiana University Press.

—— (1991b) "Cartographies of Struggle: Third World Women and the Politics of Feminism," in Mohanty *et al. Third World Women and the Politics of Feminism.*

—— (1992) "Feminist Encounters: Locating the Politics of Experience," in M. Barrett and A. Phillips (eds) *Destabilizing Theory: Contemporary Feminist Debates*, Stanford, CA: Stanford University Press.

——, Russo, A. and Torres, L. (eds) (1991) *Third World Women and the Politics of Feminism*, Bloomington, IN: Indiana University Press.

Molyneux, M. (1985) "Mobilization without Emancipation? Women's Interests, the State, and Revolution in Nicaragua," *Feminist Studies* 11, 2: 227–54.

Momsen, J. (1991) *Women and Development in the Third World*, London: Routledge.

—— (ed.) (1993) *Women and Change in the Caribbean*, London: James Curry.

—— and Kinnaird, V. (eds) (1993) *Different Places, Different Voices*, London: Routledge.

Moore, D. (1992) "The Dynamics of Development Discourse: Sustainability, Equity, and Participation in Africa," commissioned by International Development and Research Centre, Ottawa, Canada.

Moraga, C. and Anzaldua, G. (eds) (1981) *This Bridge Called My Back: Writings of Radical Women of Color*, Watertown, MA: Persephone Press. Also published (1983) New York: Kitchen Table, Women of Color Press.

Moser, C.O.N. (1989) "Gender Planning in the Third World: Meeting Practical and Strategic Gender Needs," *World Development* 17, 11: 1799–825. Reprinted in R. Grant and K. Newland (eds) (1991) *Gender and International Relations*, Bloomington, IN: Indiana University Press.

—— (1993) *Gender Planning and Development*, London: Routledge.

Mouffe, C. (1992) "Feminism, Citizenship and Radical Democratic Politics," in J. Butler and J.W. Scott (eds) *Feminists Theorize the Political*, New York and London: Routledge.

Moyo, S. (1986) "The Land Question," in I. Mandaza (ed.) *Zimbabwe: The Political Economy of Transition 1980–1986*, Dakar: Council for the Development of Economic and Social Research in Africa (CODESRIA).

Mudimbe, V. (1988) *The Invention of Africa*, Bloomington, IN: Indiana University Press.

Mueller, A. (1987) "Peasants and Professionals: The Production of Knowledge about Women in the Third World," paper presented at the Association for Women in Development meeting, Washington, DC, 15–17 April.

Mulvey, L. (1988) "Visual Pleasure and Narrative Cinema," in C. Penley (ed.) *Feminism and Film Theory*, New York: Routledge. Originally published in *Screen* 16, 3 (Autumn 1975): 6–18.

Munck, R. (1993) "Political Programmes and Development: The Transformative Potential of Social Democracy," in F.J. Schuurman (ed.) *Beyond the Impasse*, London: Zed Books.

Munday, K.E. (1992) "Human Resources Development Assistance in Canada's

Overseas Development Assistance Program: A Critical Analysis," *Canadian Journal of Development Studies* 8, 3: 385–410.

Nandy, A. (1989) "Shamans, Savages and the Wilderness: On the Audibility of Dissent and the Future of Civilizations," *Alternatives* 14: 263–77.

Nash, J. and Fernandez-Kelly, M.P. (1983) *Women, Men, and the International Division of Labor*, Albany, NY: State University of New York Press.

Nasta, S. (ed.) (1991) *Motherlands*, London: Women's Press.

Navarro, M. (1989) "The Personal is Political: Las Madres del Plaza De Mayo," in S. Eckstein (ed.) *Power and Popular Protest*, Berkeley: University of California Press.

Newland, K. (1991) "From Transnational Relationships to International Relations: Women in Development and the International Decade for Women," in R. Grant and K. Newland (eds) *Gender and International Relations*, Bloomington, IN: Indiana University Press.

Nicholson, L. (ed.) (1990) *Feminism/Postmodernism*, London: Routledge.

Nindi, B. (1990) "Experts, Donors, Ruling Elites and the African Poor: Expert Planning, Policy Formulation and Implementation – A Critique," *Journal of Eastern African Research and Development* 20: 41–67.

The Royal Norwegian Ministry of Development Cooperation (1980) "NORAD's Assistance to Women in Developing Countries," Oslo: NORAD.

Nozick, R. (1974) *Anarchy, State and Utopia*, New York: Basic Books.

Nussbaum, M. (1992) "Human Functioning and Social Justice: In Defense of Aristotelian Essentialism," *Political Theory* 20, 2: 202–46.

Nye, A. (1988) *Feminist Theory and the Philosophies of Man*, New York: Routledge, Chapman & Hall.

—— (1989) *Women and Development in Africa*, Lanham, MD: University Press of America.

Nzomo, M. (1992) *Women in Politics*, Association of African Women for Research and Development Working Paper No. 2, Nairobi: AAWORD.

—— (1993) "Political and Legal Empowerment of Women in Post-election Kenya," in M. Nzomo (ed.) *Empowering Kenyan Women*, Nairobi: National Commission on the Status of Women Publication.

O'Brien, M. (1981) *The Politics of Reproduction*, Boston: Routledge & Kegan Paul.

Odie-Ali, S. (1986) "Women in Agriculture: The Case of Guyana," *Social and Economic Studies* 35, 2: 241–90.

O'Hanlon, R. (1988) "Recovering the Subject: Subaltern Studies and Histories of Resistance in Colonial South Asia," *Modern Asian Studies* 22, 1: 189–224.

—— and Washbrook, D. (1992) "After Orientalism: Culture, Criticism, and Politics in the Third World," *Comparative Studies in Society and History* 34, 1: 141–67.

Okin, S.M. (1979) *Women in Western Political Thought*, Princeton: Princeton University Press.

Ong, A. (1988) "Colonialism and Modernity: Feminist Re-presentations of Women in Non-Western Societies," *Inscriptions* 3/4: 79–93.

—— (1990) "State versus Islam: Malay Families, Women's Bodies, and the Body Politic in Malaysia," *American Ethnologist* 17, 2: 258–76.

—— (1993) "Anthropology, China, and Modernities: The Geo-politics of Cultural Knowledge," paper presented at Association of Social Anthropology (ASA), IV Decennial Conference, Oxford, England, 26–31 July.

Opole, M. (1993) "Revalidating Women's Knowledge on Indigenous Vegetables: Implications for Policy", in W. de Boef, K. Amanor and K. Wellard with

A. Bebbington (eds), *Cultivating Knowledge*, London: Intermediate Technology Publications.

Overholt, C., Anderson, M., Cloud, K. and Austin, J. (eds) (1985) *Gender Roles in Development Projects*, West Hartford, CT: Kumarian Press.

Owens, C. (1983) "The Medusa Effect or, The Spectacular Ruse," in B. Kruger *We Won't Play Nature to Your Culture*, London: Institute of Contemporary Arts.

Pala, A.O. (1977) "Definitions of Women and Development: An African Perspective," *SIGNS* 3, 1: 9–13.

Palmer, B. (1990) *Descent into Discourse: The Reification of Language and the Writing of Social History*, Philadelphia: Temple University Press.

Palmer, I. (1991) *Gender and Population in the Adjustment of African Economies: Planning for Change*, Women, Work and Development Series No. 19, Geneva: International Labour Organization.

Palmer, R. (1990) "Land Reform in Zimbabwe, 1980–1990," *African Affairs* 8, 9: 163–81.

Pankhurst, D. (1991) "Constraints and Incentives in 'Successful' Zimbabwean Peasant Agriculture: The Interaction Between Gender and Class," *Journal of Southern African Studies* 17, 4: 611–32.

Papanek, H. (1977) "Development Planning for Women," *SIGNS* 3, 1: 14–21.

—— (1986) "Coming Out of the Niche: Intellectual Consequences of Segregating Advocacy Research on Women and Development," paper presented at Women and Development Conference, Pembroke Center for Teaching and Research on Women, Brown University, 16–17 January.

Parmar, P. (1982) "Gender, Race and Class: Asian Women in Resistance," in Center for Contemporary Cultural Studies (University of Birmingham), *The Empires Strikes Back: Race and Racism in 70's Britain*, London: Hutchinson.

Parpart, J.L. (1989) *Women and Development in Africa*, Lanham, MD: University Press of America.

—— (1993) "Who is the 'Other'?: A Postmodern Feminist Critique of Women and Development Theory and Practice," *Development and Change* 24, 3: 439–64.

—— and Staudt, K. (eds) (1989) *Women and the State in Africa*, Boulder, CO: Lynne Rienner.

Patai, D. (1988) "Constructing a Self: A Brazilian Life Story," *Feminist Studies* 14, 1 (Spring): 143–66

—— (1991) "US Academics and Third World Women: Is Ethical Research Possible?" in S. Berger Gluck and D. Patai (eds) *Women's Words: The Feminist Practice of Oral History*, New York and London: Routledge.

Pateman, C. (1980a) "The Disorder of Women, and the Sense of Justice," *Ethics* 91, 1.

—— (1980b) "Women and Consent," *Political Theory* 8, 2.

—— (1988) *The Sexual Contract*, Stanford: Stanford University Press.

—— (1989) *The Disorder of Women Democracy, Feminism and Political Theory*, Stanford: Stanford University Press.

Payer, C. (1974) *The Debt Trap: The IMF and the Third World*, Harmondsworth: Penguin Books.

—— (1982) *The World Bank: A Critical Analysis*, New York: Monthly Review Press.

Peet, R. and Watts, M. (1993) "Introduction: Development Theory and Environment in an Age of Market Triumphalism," *Economic Geography* 69, 3: 227–53.

Pescatello, A.M. (1974) "The Hispanic Caribbean Woman and the Literary Media," *Review Interamericana* 4, 2: 131–5.

Pieterse, J.N. (1991) "Dilemmas of Development Discourse: the Crisis of Developmentalism and the Comparative Method," *Development and Change* 22, 1: 5–29.

—— (1992) "Emancipations Modern and Post Modern," *Development and Change* 23,3: 5–41.

Pietila, H. and Vickers, J. (1990) *Making Women Matter: The Role of the United Nations*, London: Zed Books.

Pirages, D. and Sylvester, C. (eds) (1990) *Transformations in the Global Political Economy*, London: Macmillan.

Pires de Rio Caldeira, T. (1990) "Women, Daily Life and Politics," in E. Jelin (ed.) *Women and Social Change in Latin America*, London: Zed Books.

Poovey, M. (1988) "Feminism and Deconstruction," *Feminist Studies* 14, 1: 51–65.

Porter, D., Allen, B. and Thompson, G. (1991) *Development in Practice: Paved with Good Intentions*, London: Routledge.

Powell, D. (1986) "Caribbean Women and Their Response to Familial Experiences," *Social and Economic Studies* 35, 2: 83–130.

Prakash, G. (1990) "Writing Post-Orientalist Histories of the Third World: Perspectives from Indian Historiography," *Comparative Studies in Society and History* 32, 2: 383–408.

Pratt, B. and Boyden, J. (eds) (1988) *The Field Directors' Handbook*, Oxford: Oxford University Press.

Pratt, G. and Hanson, S. (1994) "Geography and the Construction of Difference," *Gender, Place and Culture* 1, 1: 5–30.

Pratt, M.B. (1984) "Identity: Skin Blood Heart," in E. Bulkin, M.B. Pratt and B. Smith *Yours in Struggle: Three Feminist Perspectives on Anti Semitism and Racism*, Brooklyn, New York: Long Haul Press, 9–64.

Pronk, J. (1993) *A World in Conflict*, The Hague: Government of the Netherlands.

Rabinow, P. (ed.) (1984) *The Foucault Reader*, New York, Penguin Books.

Radcliffe, S.A. and Westwood, S. (eds) (1993) *Viva: Women and Popular Protest in Latin America*, London: Routledge.

Raissiguier, C. (1994) *Becoming Women/Becoming Workers: Identity Formation in a French Vocational School*, Albany: State University of New York Press.

Rajan, R.S. (1993) *Real and Imagined Women: Gender, Culture and Postcolonialism*, London: Routledge.

Ranger, T. (1985) *The Invention of Tribalism in Zimbabwe*, Gweru: Mambo Press.

Rathgeber, E.M. (1988) "Femmes et développement: les initiatives des quelques organismes subventionnaires," *Recherches féministes* 1, 2.

—— (1990) "WID, WAD, GAD: Trends in Research and Practice," *Journal of Developing Areas* 24, 4: 489–502.

Reagon, B. (1983) "Coalition Politics: Turning the Century," in B. Smith (ed.) *Home Girls: A Black Feminist Anthology*, New York: Kitchen Table, Women of Color Press.

Redclift, M. (1987) *Sustainable Development. Exploring the Contradictions*, London: Methuen.

Reddock, R. (1985) "Women and Slavery in the Caribbean: A Feminist Perspective," *Latin American Perspectives* Issue 44, 12, 1: 63–80.

—— (1986) "Some Factors Affecting Women in the Caribbean Past and Present," in P. Ellis (ed.) *Women of the Caribbean*, London: Zed Books.

—— (1988) "Feminism and Feminist Thought: An Historical Overview," in

265

P. Mohammed and C. Shepherd (eds) *Gender in Caribbean Development*, Barbados: Women and Development Studies Project, University of the West Indies.

Review of Women's Studies (1991–2), special issue on "Women in Politics: Women, Debt and Environment," University of the Philippines, Diliman, Philippines.

Rhode, D. (ed.) (1990) *Theoretical Perspectives on Sexual Difference*, New Haven, CT: Yale University Press.

Ribe, H., Carvalho, S., Liebenthal, R., Nicholas, P. and Zuckerman, E. (1990) *How Adjustment Programs Can Help the Poor: The World Bank's Experience*, World Bank Discussion Paper No. 71, Washington, DC: World Bank.

Rich, A. (1976) *Of Woman Born*, New York: Pantheon.

Riley, D. (1988) *Am I That Name? Feminism and the Category of 'Women' in History*, Minneapolis: University of Minnesota Press.

Roe, E. (1991) "Development Narratives, Or Making the Best of Blueprint Development," *World Development* 19, 4: 287–300.

Rogers, B. (1979) *The Domestication of Women: Discrimination in Developing Societies*, New York: St. Martin's Press.

Rosenau, P. (1992) *Post-Modernism and the Social Sciences: Insights, Inroads, and Intrusions*, Princeton: Princeton University Press.

Rostow, W.W. (1960) *The Stages of Economic Growth*, Cambridge: Cambridge University Press.

—— (1984) "Development: The Political Economy of the Marshallian Long Period," in G.M. Meier and D. Seers (eds) *Pioneers in Development*, Washington, DC: World Bank.

Rothbard, M. (1977) *Power and Markets*, Kansas City: Steed Andrews & McMeel.

Saadawi, N., Mernissi, F. and Vajarathon, M. (1978) "A Critical Look at the Wellesley Conference," *Quest* 4, 2.

Safa, H. (1986) "Economic Autonomy and Sexual Equality in Caribbean Society," *Social and Economic Studies* 35, 3: 1–23.

—— (1990) "Women's Social Movements in Latin America," *Gender and Society* 4, 3 (September): 354–69.

—— and Antrobus, P. (1992) "Women and the Economic Crisis in the Caribbean," in L. Beneria and S. Feldman (eds) *Unequal Burden: Economic Crises, Persistent Poverty and Women's Work*, Boulder, CO: Westview Press.

Sahn, D.E. and Haddad, L. (1991) "The Gendered Impact of Structural Adjustment Programs in Africa: Discussion," *American Journal of Agricultural Economics* 73, 5: 1448–551.

Said, E. (1979) *Orientalism*, New York: Pantheon; also (1979) *Orientalism*, New York: Vintage Books.

—— (1993) *Culture and Imperialism*, New York: Alfred A. Knopf.

St. Hilaire, C. (1992) "Canadian Aid, Women and Development: Rebaptizing the Filipina," *Catholic Institute for International Relations* 3: 1–15.

Sakaike, T. (1991) "ZANU PF Makes the Big Switch," *Africa South* June: 8–10.

Sangari, K. and Vaid, S. (1989) *Recasting Women: Essays in Colonial History*, New Delhi: Kali Press.

Sargent, L. (ed.) (1981) *Women and Revolution: A Discussion of the Unhappy Marriage of Marxism and Feminism*, Boston: South End Press.

Sawicki, J. (1991) *Disciplining Foucault*, New York: Routledge.

Sayer, D. (1991) *Capitalism and Modernity*, London: Routledge.

Schick, I.C. (1990) "Representing Middle Eastern Women: Feminism and Colonial Discourse," *Feminist Studies* 16, 2: 345–80.

Schmidt, E. (1990) "Negotiated Spaces and Contested Terrain: Men, Women, and the Law in Colonial Zimbabwe, 1890–1939," *Journal of Southern African Studies* 16, 4: 622–48.

Schmink, M. (1981) "Women in Brazilian *Abertura* Politics," *SIGNS* 7, 11: 115–34.

Schumacher, E.F. (1973) *Small Is Beautiful*, London: Blond & Briggs.

Schuurman, F.J. (ed.) (1993a) *Beyond the Impasse*, London: Zed Books.

—— (1993b) "Introduction: Development Theory in the 1990s," in F.J. Schuurman (ed.) *Beyond the Impasse*, London: Zed Books.

—— (1993c) "Modernity, Post-modernity and the New Social Movements," in F.J. Schuurman (ed.) *Beyond the Impasse*, London: Zed Books.

Scott, J.C. (1985) *Weapons of the Week: Everyday Forms of Peasant Resistance*, New Haven: Yale University Press.

Scott, J.W. (1988) "Deconstructing Equality – versus Difference: Or the Use of Poststructuralist Theory of Feminism," *Feminist Studies* 14, 1: 33–50.

—— (1992) "'Experience'," in J. Butler and J.W. Scott (eds) *Feminists Theorize the Political*, New York and London: Routledge.

Sen, G. and Grown, C. (1987) *Development, Crises, and Alternative Visions: Third World Women's Perspectives*, New York: Monthly Review Press.

Shapiro, M. (1981) *Language and Political Understanding: The Politics of Discursive Practices*, New Haven: Yale University Press.

Sharpe, J. (1991) "The Unspeakable Limits of Rape: Colonial Violence and Counter-Insurgency," *Genders* 10: 25–46.

Shiva, V. (1988) *Staying Alive: Women, Ecology and Development*, London and Delhi: Zed Books and Kali for Women.

Shohat, E. (1992) "Notes on the 'Post-Colonial'," *Social Text* 31/32: 99–113.

Silman, J. (ed.) (1987) *Enough is Enough: Aboriginal Women Speak Out*, Toronto: Women's Press.

Slater, D. (1992) "Theories of Development and Politics of the Post-modern," *Development and Change* 23, 3: 283–319

—— (1993) "The Political Meanings of Development," in F.J. Schuurman (ed.) *Beyond the Impasse*, London: Zed Books.

Smith, B. (ed.) (1983) *Home Girls: A Black Feminist Anthology*, New York: Kitchen Table, Women of Color Press.

Smith, D.E. (1987a) "Women's Perspectives as a Radical Critique of Sociology," in S. Harding (ed.) *Feminism and Methodology*, Bloomington, IN: Indiana University Press and Open University Press.

—— (1987b) *The Everyday World as Problematic: A Feminist Sociology*, Toronto: University of Toronto Press.

—— (1988) "Femininity as Discourse," in L. Roman, L. Christian Smith, and E. Ellsworth (eds) *Becoming Feminine: The Politics of Popular Culture*, London, New York: Falmer Press.

—— (1990) *The Conceptual Practices of Power*, Toronto: University of Toronto Press.

Smith, M.G. (1965) *The Plural Society in the British West Indies*, Berkeley: University of California Press.

Smith, R. (1988) *Salt on the Snow*, London: Bodley Head.

Smith, R.T. (1956) *The Negro Family in British Guiana*, London: Routledge & Kegan Paul.

Snyder, M. (1992) Private Communication.

Spelman, E. (1990) *Inessential Woman*, London: Woman's Press.

Spivak, G.C. (1987) *In Other Worlds: Essays in Cultural Politics*, New York: Methuen, and (1988) New York: Routledge.

――― (1988) "Can the Subaltern Speak?" in C. Nelson and L. Grossberg (eds) *Marxism and the Interpretation of Culture*, Urbana: University of Illinois.

――― (1990) *The Post-Colonial Critic: Interviews, Strategies, Dialogue*, edited by S. Harasym, London: Routledge.

Stamp, P. (1989) *Technology, Gender and Power in Africa*, Technical Study 63E, Ottawa: International Development Research Centre.

Staudt, K. (1985) *Women, Foreign Assistance and Advocacy Administration*, New York: Praeger.

――― (1986) "Women, Development and the State: On the Theoretical Impasse," *Development and Change* 17: 325–33.

Steady, F.C. (1987) "African Feminism: A Worldwide Perspective," in R. Terborg-Penn, S. Harley and A.B. Rushing (eds) *Women in Africa and the African Diaspora*, Washington, DC: Howard University Press.

――― (1990) "Women: The Gender Factor in the African Social Situation," in *The African Social Situation: Crucial Factors of Development and Transformation*, published for the African Centre for Applied Research and Training in Social Development (ACARTSOD), London: Hans Zell.

Stein, S. and Stein, B. (1970) *The Colonial Heritage of Latin America*, Oxford: Oxford University Press.

Stichter, S. and Parpart, J. (eds) (1988) *Patriarchy and Class: African Women in the Home and the Workplace*, Boulder, CO: Westview Press.

――― (eds) (1990) *Women, Employment and the Family in the International Division of Labour*, London: Macmillan.

Stubbs, R. and Underhill, G. (eds) (1994) *Political Economy and the Changing Global Order*, Toronto, McClelland & Stewart.

Suleri, S. (1992) *The Rhetoric of English India*, Chicago: Chicago University Press.

Sutton, F.X. (1963) "Social Theory and Comparative Politics," in H. Eckstein and D. Apter (eds) *Comparative Politics: A Reader*, New York: Free Press.

Sylvester, C. (1986) "Zimbabwe's 1985 Elections: A Search for National Mythology," *Journal of Modern African Studies* 24, 2: 229–56.

――― (1990a) "Unities and Disunities in Zimbabwe's 1990 Election," *Journal of Modern African Studies* 28, 3: 375–400.

――― (1990b) "Simultaneous Revolutions: The Zimbabwean Case," *Journal of Southern African Studies* 16, 3: 452–75.

――― (1991a) "'Urban Women Cooperators', 'Progress', and 'African Feminism' in Zimbabwe," *Differences* 3, 1: 9–62.

――― (1991b) *Zimbabwe: The Terrain of Contradictory Development*, Boulder, CO: Westview Press.

――― (1994) *Feminist Theory and International Relations in a Postmodern Era*, Cambridge: Cambridge University Press.

Tinker, I. (1990) "The Making of a Field: Advocates, Practitioners and Scholars," in I. Tinker (ed.) *Persistent Inequalities: Women and World Development*, Oxford: Oxford University Press.

――― and Bramsen, M. (1976) *Women and World Development*, Washington, DC: Overseas Development Council.

Torres, L. (1991) "The Construction of the Self in US Latina Autobiographies," in C. Mohanty, A. Russo and L. Torres (eds) (1991) *Third World Women and the Politics of Feminism*, Bloomington, IN: Indiana University Press.

Tress, D.M. (1988) "Comments on Flax's 'Postmodernism and Gender Relations in Feminist Theory'," *SIGNS* 14, 1: 196–200.

United Nations (1985) The Nairobi Forward Looking Strategies for the Advancement of Women, Conference to Review and Appraise the Achievements of the United Nations Decade for Women: Equality, Development and Peace, Nairobi, Kenya, 15–26 July, New York: United Nations.

United Nations, Department of International Economic and Social Affairs (1986) *World Survey on the Role of Women in Development*, New York: United Nations.

United Nations Development Programme (UNDP) (1993) *Human Development Report*, New York: United Nations.

United States Agency for International Development (USAID) (1982) "Women in Development," USAID Policy Paper, Washington, DC: USAID.

———— (1987) *Women in Development: AID's Experience, 1973–1985*, Vol. 1, Synthesis Paper, USAID Program Evaluation Report No. 18, Washington, DC: USAID.

Valenzuela, J.S. and Valenzuela, A. (1981) "Modernization and Dependency: Alternative Perspectives in the Study of Latin American Underdevelopment," in H. Muñoz (ed.) *From Dependency to Development: Strategies to Overcome Underdevelopment and Inequality*, Boulder, CO: Westview Press.

Van Kirk, S. (1980) *"Many Tender Ties": Women in Fur-Trade Society in Western Canada, 1670–1870*, Winnipeg: Watson & Dwyyer.

Vargas, V. (1992) "The Feminist Movement in Latin America: Between Hope and Disenchantment," *Development and Change* 23, 3: 195–214.

Verhagen, K. (1987) *Self-Help Promotion: A Challenge to the NGO Community*, Amsterdam: Cebemo.

Walby, S. (1990) "A Critique of Postmodernist Accounts of Gender," paper presented at Canadian Sociological Association Meetings, Vancouver, British Columbia.

Walker, A. (1983) *In Search of My Mother's Garden*, New York: Harcourt Brace Jovanovich.

Ward, K. (1986) *Women and Transnational Corporation Employment: A World Systems and Feminist Analysis*, Women in Development Working Paper No. 120, East Lansing, MI: Office of Women in International Development, Michigan State University.

———— (ed.) (1990) *Women Workers and Global Restructuring*, Ithaca, NY: Cornell University Press.

Watson, L.C. (1976) "Understanding a Life History as a Subjective Document," *Ethos* 4: 95–131.

———— and Watson-Franke, M.-B. (1985) *Interpreting Life Histories: An Anthropological Inquiry*, New Brunswick, NJ: Rutgers University Press.

Watts, M. (1993) "Development I: Power, Knowledge, Discursive Practice," *Progress in Human Geography* 17, 2: 257–72.

Weedon, C. (1987) *Feminist Practice and Poststructuralist Theory*, Oxford: Basil Blackwell.

Weinrich, A.K.H. (1979) *Women and Racial Discrimination in Rhodesia*, Paris: UNESCO.

Weiss, R. (1986) *The Women of Zimbabwe*, London: Kesho.

Wellesley Editorial Committee (1977) *Women and National Development: The Complexities of Change*, Chicago: University of Chicago Press.

Wexler, P. (1987) *Social Analysis of Education: After the New Sociology*, London and New York: Routledge & Kegan Paul.

White, A. (1986) "Profiles of Women in the Caribbean Project," *Social and Economic Studies* 35, 2: 59–82.

White, S. (1986) "Foucault's Challenge to Critical Theory" *American Science Review*, 80.

Whitehead, A. (1985) "Effects of Technological Change on Rural Women: A Review of Analysis and Concepts," in I. Ahmed (ed.) *Technology and Rural Women: Conceptual and Empirical Issues*, London: George Allen & Unwin.

Wilson, A. (1985) *Finding a Voice: Asian Women in Britain*, London: Virago.

Wiltshire, R. (1988) "Indigenisation Issues in Women and Development Studies in the Caribbean: Towards a Holistic Approach" (mimeo).

Women and Development Studies Unit (1990) *Women Speak: A Magazine About Caribbean Women*, Nos. 26 and 27, Barbados: University of the West Indies.

Wood, G. (1985) "The Politics of Development Policy Labelling," *Development and Change* 16: 347–73.

Wood, R.E. (1986) *From Marshall Plan to Debt Crisis: Foreign Aid and Development Choices in the World Economy*, Berkeley: University of California Press.

World Bank (1979a) *Recognizing the "Invisible" Woman in Development*, Washington, DC: World Bank.

—— (1979b) *World Development Report*, Washington, DC: World Bank.

—— (1989a) *Sub-Saharan Africa: From Crisis to Sustainable Growth*, Washington, DC: World Bank.

—— (1989b) *Kenya World Bank Country Study Report*, Washington, DC: World Bank.

—— (1989c) *Women in Development: Issues for Economic and Sector Analysis*, Policy, Planning and Research Working Paper No. 269, Washington, DC: World Bank.

—— (1990) *Women in Development: A Progress Report on the World Bank Initiative*, Washington, DC: World Bank.

—— (1993) "Paradigm Postponed: Gender and Economic Adjustment in Sub-Saharan Africa," Technical Note, Human Resources and Poverty Division, Technical Department, Africa Region, Washington, DC: World Bank.

Young, I. (1981) "Beyond the Unhappy Marriage: A Critique of the Dual Systems Theory," in L. Sargent (ed.) *Women and Revolution: A Discussion of the Unhappy Marriage of Marxism and Feminism*, Boston: South End Press.

Young, K., Walkowitz, C. and McCallogh, R. (eds) (1981) *Of Marriage and the Market*, Berkeley: University of California Press.

—— (1988) "Reflections on Meeting Women's Needs," in K. Young (ed.) *Women and Economic Development*, New York: Berg/UNESCO.

Young, K. (1988) "Notes on the Social Relations of Gender," in P. Mohammed and C. Shepherd (eds) *Gender in Caribbean Development*, Barbados: Women and Development Studies Project, University of the West Indies.

Young, R. (1990) *White Mythologies*, London: Routledge.

ZANU PF Election Manifesto 1990, Harare: Jongwe.

Zeroulou, Z. (1987) "A L'Ile et dans la Région du Nord: la Seconde génération entre en faculté," *Hommes et Migrations* 1108.

—— (1988) "Familles Immigrées et 'Ecole Française': Quels types de rapports," *Migrants-Formation* December, 21–6.

Zimbabwe, Government of (1982) *Transitional National Development Plan: 1982/83–1984/85*, Vol. 1, Harare: Government Printer.

—— Ministry of Community and Cooperative Development and Women's Affairs (1983) *Survey on Women*, Harare: Government Printer.

INDEX